INTERNATIONAL TRANSPORT

INTERNATIONAL TRANSPORT

An Introduction to Current Practices and Future Trends

Rex W. Faulks

YOURS TO HAVE AND TO HOLD

BUT NOT TO COPY

First published as *Elements of Transport* by Ian Allen Ltd in 1965. Second edition of *Elements of Transport* published in 1969. Published as *Principles of Transport* in 1973, 1977 and 1982 by Ian Allen Ltd and again in 1990 by McGraw-Hill Book Company (UK) Limited.

First published as *International Transport* by Kogan Page Limited in 1999.

Kogan Page Limited
120 Pentonville Road
London N1 9JN
UK

CRC Press
Rights and Permissions
2000 NW Corporate Blvd
Boca Raton
Florida 3343
USA

© Rex W. Faulks, 1999

British Library Cataloguing in Publication Data

A CIP record for this book is available from the British Library.

ISBN 0 7494 2832 5

Library of Congress Catalog Card Number 99-72248
ISBN 0 8493 4083 7

Typeset by Saxon Graphics Ltd, Derby
Printed and bound in Great Britain by Biddles Ltd, Guildford and King's Lynn

Contents

Preface

International Transport is the successor book to *Principles of Transport* which, together with its predecessor *Elements of Transport* – first published in 1965 – has had a successful run through six editions. The subject matter has relevance to the requirements of students of the Chartered Institute of Transport as it deals in general terms with the movement of people and goods which are important parts of the syllabus of the Institute's Diploma in Transport. It is aimed at the needs of school leavers and undergraduates without previous experience but who intend to make a career in transport and at anyone else who is new to the study of transport. The first five chapters in particular are written in a simple style and represent the basics of the elements of transport.

The relevant information contained in the previous editions has been brought forward suitably updated. In this edition the practice of taking a quick look into the future has been retained and enlarged. Eight years have elapsed since the previous book was written and during that period much has changed. Not only have the Institute's syllabuses been revised, which means that the needs of the students are different, it has also become increasingly apparent that transport, and particularly the movement of people, is on a collision course with the availability of space to move around and also with the preservation of the environment. In looking into the future, the author views with alarm the situation as he considers it will become, for example, on the roads in urban areas, if the growth of private transport continues unchecked. Furthermore, the impact of transport on the environment could be devastating if things are not properly handled and this means worldwide including both the developed and the developing countries. There is now so much talk on this subject (although very little being done) that no generalized book on transport would be complete without reference to it. It is for this reason that thoughts on the future cannot now be written off in a few pages and this part of the book has therefore been expanded.

Whilst what follows constitutes a review of the fundamental facts and issues relative to transport in all its forms, it is inevitable that the final chapters which look into the future contain subjective thought. Possible courses of action which are either suggested or implied may be at variance with what has been said by other people in other places. However, this does not imply criticism. Where this arises it reflects a different point of view with different priorities in mind. Any thoughts on the environment that have been presented are submitted with regard only to transport issues. They are formulated in the 'ideal' sense and in the cold light of natural evolution and development which remain aloof from the humanities. The imposition of thoughts, opinions and suggestions, particularly when originating from more than one source, stimulates mental argument. In any case, one of the reasons for including ideas on possible courses of action in the future is to provoke discussion and for this reason they are deliberately controversial. It is necessary that students learn the facts, listen to the sentiments of others and then develop their own ideas. Where thoughts have been expressed there is an indication as such as it is important that students do not mix fact with opinion. Some of the proposals here are inevitably harsh and would be unpopular if implemented. The critics will have their day (and their say!).

By style and presentation it is the masculine pronoun which is used throughout. British readers may say that reference to 'his' violates the terms of the (British) Sex Discrimination Act, 1975. For their benefit, may it be said that the (British) Interpretation Act, 1889 stipulates that in statutory documents the term 'he' will be deemed to embrace 'he' or 'she' where appropriate. It is, therefore, regarded here as unnecessarily pedantic to elucidate further on pronouns of what are in this context recognized as an agreed common gender. As the British parliament seems prepared to accept this, the writer also uses the masculine term which by inference includes the feminine as well and certainly no disparagement of the female reader is either implied or intended.

Beyond what has just been said, no direct references to statutes are contained in this work. The book is intended for the international market and the legal complexities differ nation by nation. To incorporate all of the detail and the infinite number of variations would be impracticable and in any case unnecessary as transport law extends well beyond the basic principles. References to legal matters are, therefore, made only in descriptive terms.

In presenting this book it is hoped that its pages will offer guidance and interest. Every effort has been made to ensure accuracy and that no misrepresentation has occurred.

Acknowledgements

As in previous editions, some of the information and also the background to some of the case studies has been supplied by various organizations associated with the transport industry. The author gratefully acknowledges such help which has been freely given by the following bodies:

BAA (British airports)
Baltic Exchange
Brisbane Transport
British Nuclear Fuels
British Road Federation
Bureau of Transport and Communications Economics, Canberra
Canadian Minister of Transport
Chartered Institute of Transport
Electricity Association
Essener Verkehrs-AG (Essen Transport)
European Commission (Directorate-General for Transport)
European Federation for Transport and the Environment
Greenergy International
Hoverspeed
International Air Transport Association
International Chamber of Shipping
International Civil Aviation Organization
International Maritime Organization
International Road Federation
International Road Transport Union
International Union (Association) of Public Transport
International Union of Railways

Japanese Embassy
Lloyd's Register
London Transport
Metropolitan Bus Service, Johannesburg
Metro Transit, Seattle
National Power
National Wind Power
New Zealand Minister of Transport
Peninsular and Oriental Steam Navigation Co (P&O)
Port of London Authority
RAPT (Paris Transport)
State Transit Authority, New South Wales
Stena Line
STIB (Brussels Transport)
Toronto Transit Commission
UK Atomic Energy Authority
United Nations Environment Programme
World Health Organization

In some cases photographs have been supplied and where this has been done specific acknowledgements accompany the illustrations.

Introduction

This is an introduction to a subject which in itself is an introduction. It is all-embracing in the widest sense and incorporates a multitude of different skills, systems and services in as many different countries worldwide. The different transport modes have their different applications and the common core is the function to move people or goods from where they are to where they would prefer to be or to where their relative values are greater.

In recent years there has been an added duty and one common to every member of society: to preserve the environment. But in the world of transport there are opposing forces. To refrain from polluting the atmosphere and to eliminate noise and visual intrusion whilst at the same time continuing to provide the services which are essential to everyday life puts transport provision in a situation where the requirements are mutually exclusive. But if nothing is done, the prophesied results are serious. Certainly, transport is only one function that is damaging the environment but it is an important one and for this reason it is necessary to give special attention to this aspect. A concomitant problem is space (or the lack of it) and that means space for people to live, work and move around. The effects of all of these situations are aggravated if the world population increases in what is a static or even a diminishing availability of land space.

At one time the 'transporters' provided their services as part of a commercial business and in so doing gave an essential service to the community. For the conveyance of merchandise they still do but on parts of the passenger side much of the profit has evaporated whilst the need remains. Although many onetime commercial passenger services have now become social services there is also a swing worldwide towards privatization. This has brought with it the problem of how to support private industry from public funds. But with monetary support from outside to bolster the provision of social services it is easier to exercise some form of control and hence pay more regard to the systems concept. Without the personal profit motive, operators are better able to yield to the proposals of independent authorities: to co-ordinate rather than compete. In this way the characteristics of each mode can be properly exploited and wasteful competition avoided. This, of course, is only one theory. The opposite concept is that competition is in itself the best spur to efficiency and that in the end this is the policy which will produce the most attractive services at the lowest cost. At this stage the issue becomes political. But political science and transport are two distinct and very different subjects and it is transport which is under discussion here. It is unfortunate if political dogma is allowed to overshadow and hence dictate policies which are not necessarily in the best interests of professional transport.

An attempt has been made to draw together the fundamental issues which are of common importance in whatever branch of transport the reader may be involved. It is a springboard for subsequent, more specialized studies. What, however, must be appreciated, and this is most important if the book is to have any credence at all, is the enormity of the subject that has been tackled. In comparative terms, the preparation involved has resembled compressing a five-storey tenement block into something akin to the size of an ice cube. Much detail has of necessity been omitted and even the stated principles are in a skeletal form only. There can therefore be little scientific examination of the peculiar complexities of any one mode. Following this condensed version of the subject, students will need to go on to assimilate more advanced texts. For environmental matters, the 1994 report of the (British) Royal Commission on Environmental Pollution is particularly recommended. Otherwise, they will benefit from the many specialized papers presented by experts through professional bodies and learned societies and they may perhaps participate in seminars. A study of the principles of transport as a subject is essentially a prelude and in no way must it be regarded as a substitute for those detailed and specific writings. In any case, things change. They are changing all the time and in some cases with such rapidity that as far as textbooks are concerned, in parts they may quickly become dated. In view of these constantly changing patterns, it is necessary to keep abreast of developments by reading the contemporary transport press where technical articles are written by modal specialists. The quality of their information and presentation is high but, when relating to a general overview (and no criticism is intended here), allowance must be made for any singular approach that pays little regard for the systems concept.

Different modes each have their particular skills and the specialists cannot be expert in them all. This book draws together the basics of all of the modes through which the principles have a common thread and offers a foundation for independent judgement. The subject is treated internationally and, apart from examples and case studies, excludes detail relative to any one country. The examples and case studies are, however, drawn from a range of countries across the world. If the pages which follow are seen to cover the fundamentals of transport in the widest interpretation in a potted form which can be accepted as a pointer to further study, then the purpose will have been achieved.

R.W.F.
September 1998

ONE

Systems and Strategies

BACKGROUND

This is not a historical treatise. Much has already been written on the development of transport through the ages and for this particular study it is neither opportune nor necessary to once again discuss the evolution of the wheel and how primitive man first made his incursions into the domains of neighbouring tribes. Nevertheless, in any study of transport, what has happened in the past cannot be ignored if there is to be a proper understanding of the present transport systems and the facilities that they provide, together with the many resulting transport issues. Some brief historical references are therefore necessary.

Transport services were in the main pioneered by private enterprise and they sprang from humble beginnings. There were exceptions – for example, in parts of northern Europe and in New Zealand where railway construction was financed from public funds – but this was not generally the case. Then, around the turn of the nineteenth century, municipalization became the order of the day when local authorities found themselves the owners and operators of many hitherto privately owned and developed local tram, bus and ferry services. Halfway through the twentieth century nationalization became fashionable when large road, railway and airline companies passed into state ownership with the resultant mammoth organizations. Now there is a worldwide trend back to private ownership, in many cases with large-scale enterprise broken down into smaller competitive units. This is a principle which represents current thinking and is often seen as the panacea for all transport ills. A background knowledge of the evolvement of the different types of operating units, their ownership, size and control is therefore desirable.

The environment is now one of the most talked-about issues of the day. After years of haphazard growth with inadequate attention afforded to building development, including transport infrastructure, and pollution in its various forms, people generally (but particularly in the developed countries) have become increasingly aware of the resultant ill effects that are now being witnessed across the world. Hence, and again mainly in the developed world, there is a growing awareness of the need to moderate or prevent what in the longer term could be dire effects. Pressure groups are particularly vocal on the preservation of the environment, but politically it is a delicate issue. Worldwide, national governments pay lip service and nibble at the subject, but to date there has been little concerted remedial action. Nevertheless, transport is currently under criticism for its contribution to what is now recognized as a deteriorating environment and in view of the importance of the various issues, Chapter 8 is devoted to this particular subject.

This chapter is introductory. Its purpose is to set the scene and indicate the more important issues which affect the transport industry today. Whilst some of these topics will be elaborated upon here, a more detailed examination of others will be reserved for discussion later.

DEVELOPMENT

On both land and water, propulsion was originally by natural power – wind on the sea and the exertion of humans and animals on land and along rivers and canals (some canals were constructed at a very early stage). In London (England), for example, in the seventeenth and eighteenth centuries, wherries, being light shallow passenger-carrying rowing boats, provided one of the fastest means of local transport. Bearing in mind that even by 1800 no part of London was more than a mile (1.6km) from the river it was often quicker to take a boat and then walk the rest of the way. (Interestingly, some 200 years on, we are still being invited to walk, this time in the interests of pollution.) A water taxi was summoned by standing on the bank and shouting 'oars' – a performance which was promptly followed by a scramble for the potential customer; hence an early example of small-scale private competitive enterprise. Then came the steam engine and with it the steamship and the railway.

But to stop there for a moment: the first form of conveyance for goods and passengers (other than with pack animals) was on the water with wind and sail. It was for this reason that people began to cluster around the areas best able to be served by those waterborne vessels which made possible their international and domestic communications. In other words, they settled at physically accessible and acceptable places along coastlines and astride navigable rivers, and it was here that traders first began to conduct their business. The site of London was fixed by the Romans nearly 2000 years ago because it was the lowest point on the river where they could construct a bridge and it was also a point to where their ships could navigate. Today many of the world's major centres of population and commerce are so situated because their roots date back to the transport requirements of the time; so much so that those that are not can be readily identified. Brasilia and Canberra, the capital cities of Brazil and Australia respectively, are examples to the contrary. But these two townships were founded more recently when road, rail and particularly air transport could easily be provided and which in the event happened. Interestingly, however, unlike the older established centres of those two countries, the airports of these two more modern cities have few if any international flights.

Right from the beginning of man's waterborne trading and travelling activities, therefore, transport was fixing the location of the conurbations of the future. It has been responsible for the ongoing process of setting the scene, drawing the map, shaping the skyline and hence influencing the development of the environment – an evolutionary process which has embraced the entire surface of the earth.

It was the latter part of the eighteenth century when the coal-burning steam engine (sometimes wood) was first developed in Great Britain. Although not originally designed for transport purposes, it was soon to be applied to steamships and made possible the birth of the railway; it was also tried on the roads but with only limited success. The use of steam brought about worldwide networks of shipping services and railways to be followed in due course by the use of oil-based fuels, first, around the turn of the nineteenth century, on the roads and then on rail and sea and finally in the air. Slightly earlier than the inception of the internal combustion engine came also the production of electric power which could be fed into fixed track systems (ie, railways and tramways) for propulsion purposes and which developed alongside the other modes. The evolution of transport from the steam-operated railway trains to passenger-carrying aircraft spanned a period of about a hundred years and services were built up in accord with the technological developments of the day. Steamship, canal-carrying, railway, tram, bus, road haulage and airline companies came into being over the years to offer new facilities, usually to produce a return on capital invested by optimistic speculators. Some of these new facilities met a need and some created their own need as they stimulated economic growth.

Transport remained instrumental in opening up new territories. For example, in the early 1900s when the Underground railway in London (England) extended its tentacles into the green fields surrounding the metropolis, the people followed and rural countryside was transformed into dormitory areas which are now integral parts of Greater London. The same pattern was followed, be it by train or tram, in many of the other original settlements worldwide. Some 30 years later the introduction of scheduled air services enabled hitherto remote areas such as Central and East Africa to develop and the pioneering work of the Queensland and Northern Territory Aerial Services (now known as QANTAS) opened up the Australian outback. There are countless other similar examples in every continent of the world where transport has been responsible for development and consequent change in the character of the environment, not least being the creation of demand by the cheap air charter market with its development of tourism and all the ramifications of that. It is only in more recent time that there has been expansion in rural undeveloped areas not already served by public transport (new towns catering for population overspill, for example) but where, like the Brasilia and Canberra examples, it was known that adequate communication was a physical possibility and that the requirements would be met. Hence there has been a transition from transport catering for demand by its own volition and in the process changing the environment out of all recognition, to transport chasing a demand which has been created by the activities of man in his quest for space to live, work and have his recreation. Today, road construction is very much a part of the latter scenario but it is also a highly contentious environmental issue.

As the populations grew larger and the economies grew stronger so public transport services were extended until they overlapped and duplicated each other. That is why in Great Britain where railways were first introduced and which, in terms of square miles, is a relatively small closely knit country, many of the larger provincial towns came to be served by two railway companies with separate stations providing parallel facilities. This meant that services for common destinations were fragmented and passenger interchange was difficult. It was also uneconomic. This is a result of development taking place through speculative and haphazard growth rather than to a basic preconceived plan. To take a specific example, the erstwhile Great Central and Midland Railway Companies in Great Britain had parallel routes between Sheffield and London, a distance of some 160 miles (260km), serving the same intermediate main towns on the way.

In the western world, public transport on road and rail reached its zenith around the 1950s when movement in relative comfort, speed and safety had become a reality. Wartime conditions had retarded the growth of the private car, petrol was still scarce and air transport with its distraction of cheap charter trips overseas had not begun. Nevertheless, after years of deprivation and pent-up desires to get out and about, people now clamoured to go places and see things. The result was a strong demand for domestic public transport and a wide range of services had come to be provided. Although these circumstances coincided with the nationalization era with its large-scale operating units tending to bring with them higher labour costs, they came before much of the present industrial, social and safety legislation, also with its impact on costs. Under the conditions which prevailed at that time, coupled with the ready demand for public transport, it was still possible to provide comprehensive facilities at affordable fares. But the situation was not to last. Competition from private transport came; in much of the developed world costly industrial legislation came and, together with the mammoth operating combines, overhead expenses increased. Heavy deficits began to appear, particularly on the railways and rural buses, and subsidies became necessary to support unremunerative but socially desirable services. Then the air transport industry, still young but with a fund of experience gained from wartime combat activities, began to develop and large-scale movement of passengers by air was becoming a reality, although in those early developing years financial support for both airlines and their terminals was required. Money from public funds was therefore sought by road, rail and air transport (all mainly passenger) which meant political involvement. Throughout the subsequent years there have been deliberations by governments throughout the world on the merits or otherwise of supporting public transport from public funds and the methods and extent to which that support should be given. Their arguments have overspilt into reorganizations and regulations (or de-regulations), all being coloured by their particular political inclinations. Hence came the great debate on the relative merits of large or small units, co-ordination or competition, and now the interests of the environment, whilst at the same time becoming entangled with the worldwide phenomenon of privatization throughout industry generally.

In all of this turmoil and whatever the circumstances and interventions from without, transport, although a political shuttlecock, must remain a public service and collectively it must continue to meet the prevailing demand for the movement of people and merchandise. Regardless of the mode, the prime objective of the transport industry must be a satisfactory and safe arrival at an acceptable cost.

MOTIVE POWER

It has been seen that at one time transport relied on natural power for propulsion – sea transport on the wind and animals on shore. But the advent of the steam engine and later the internal combustion engine revolutionized methods of propulsion to the extent that there is now no place in either private or commercial transport for power in its natural state (other than for bicycles). Nevertheless, all forms of power emanate from natural sources. Coal, however, has now declined in importance and where steam propulsion remains, the boilers of some railway locomotives (of those that are left) and ships are more conveniently fired by oil.

The raw material of artificially produced power which is carried on the unit itself is now largely oil-based with diesel, petrol or jet engines as the case may be. There has also been a limited use of road vehicles independently powered by electricity from charged batteries. Work which involves only a limited range without a need for high speeds but with frequent stops and starts has lent itself to this type of traction, the classic example being the vehicles used for household milk delivery in Great Britain. Otherwise, investigations into a more robust and large-scale adoption of this system has been ongoing for many years; there has been progress but it is still the subject of research and development. For electrically powered fixed-track systems the fuel is of course not so carried. Electric power, the sources of which include coal, oil, water power (hydro-electric) and nuclear fission, is produced at remote points and fed into the fixed-track systems. The artifical way then becomes also a powered way.

The distinction between an oil-powered engine carried on the unit and a fixed-track system with power fed to it, having been generated at a remote point, is fundamental in terms of both cost and the environment. Cost is involved because the capital necessary to equip an electric railway or tramway is substantial, hence the traffic potential must be sufficient to justify such expenditure. In environmental terms there is a repercussion because an electric vehicle does not itself pollute the atmosphere although the power does have to be generated (at a remote point) which may cause pollution.

The form of motive power used may also be influenced by local conditions. In some countries oil may be plentiful and cheap (although this may not last); in others oil may not be readily obtainable but there might be an abundance of water power waiting to be harnessed. Then there are the artificial reasons. A certain type of fuel might be subject to heavy taxation. A transport company might also be the electric supply company and examples of this have been known in the case of street tramways. Considerations such as these have influenced the choice of power.

What has just been said constitutes a brief summary of the different types of motive power used for transport purposes since the days of the horse. But they are not all environmentally friendly and in this era of environmental susceptibility, their continued use is of no small concern. The matter is, therefore, now the subject of much research and the development of cleaner types of fuel (liquid and gas) is an ongoing process. It is appropriate, therefore, to continue the examination of this subject in Chapter 8.

PUBLIC AND PRIVATE TRANSPORT

'Public' and 'private' are two basic concepts of transport provision and they have no connection with public or private ownership as is discussed in Chapter 4. In road haulage, for example, any business, be it a mammoth combine or a small, sole proprietor carrying its own goods in its own vehicles (be they small vans or 44 tonne trucks) as a part of its manufacturing or trading activities (ie, the 'ancillary' or 'own account' operators) constitutes private transport. This would be regardless of whether the organization concerned happens to be in the public or private sector, in terms of ownership. The professional carrier whose transport business is to carry other people's goods constitutes public transport and gives a 'hire and reward' service. On the passenger side the distinction is more obvious. Public passenger transport on the roads is the bus even though the vehicles are likely to be privately owned, whilst private transport is the private car. Similar distinctions apply with the other modes, particularly on the goods side. Compare, for example, the general purpose bulk carriers of shipping lines with the oil tankers owned by one of the major oil companies.

The main advantages and disadvantages of private transport to the user in the carriage of merchandise may be summarized as follows:

- Management has better control over the transport staff as they are its own employees.
- Management has sole use of its fleet and can plan its operations to the best advantage.

But

- The operation of a private transport fleet is an added responsibility. Unless the work is sufficient to justify a specialist transport department it might be better left in the hands of professional carriers.
- A company must bear not only the direct running costs but also the entire overhead expenses of a private fleet. Work must therefore be sufficient to keep the vehicles or vessels constantly employed. Public transport is paid for as it is used.

It is on the passenger side that mention of private transport brings with it one of the greatest transport problems of the day and that is on the roads with the proliferation of the private car. Private motoring has become inordinately popular for both business and pleasure purposes. Experience so far has shown that no amount of added fuel tax, insurance cost, parking charges or any other associated expenses have detracted from its popularity (although the size and scope of these deterrents have not yet been exhausted). As a part of the transport system the car reigns supreme and the phenomenon is worldwide. Figure 1.1 illustrates the growth of car ownership in Great Britain over the years from 1950, with future projections (other things being equal). It is not practicable nor particularly useful to produce worldwide figures and those presented are in respect of only one country. Nevertheless, Figure 1.1 shows a pattern which is reasonably representative of the Western world. Other figures prepared by the European Commission in respect of its member countries and which are mentioned below under the sub-section 'Published planning proposals' show a markedly similar trend which confirms this view.

From 1950 (the height of the public transport boom) to 1996, car ownership in Great Britain increased elevenfold. For every single car that was on the road then there are 11–12 today. Over the ten years 1986 to 1996 there was an increase of some 25 per cent. Looking at the top forecast of

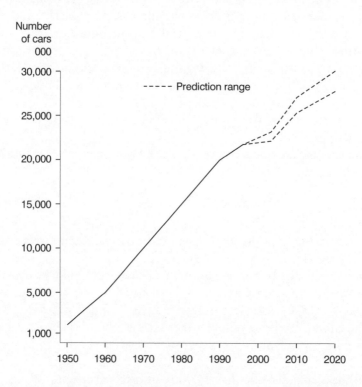

Figure 1.1 Graphical indication of private car ownership in Great Britain.

Source: Transport Statistics, Department of Transport (GB)

the predicted range for the future, the 1996 figure could increase by approaching 30 per cent in 2010 and by nearly 45 per cent by 2020. In this scenario there would be nearly half as many more cars on the road in 2020 as there are today – except that in the built-up areas there would not be room on the roads for them. Neither would there be space or, to some people, the inclination to build additional roads.

It is acknowledged that the ownership of personal transport has always been, still is, and probably always will be regarded as an enhancement and now an essential part of the quality of life. It is something that everybody wants and it must, therefore, be accepted that private cars must not be artificially restricted without good reason. Unfortunately, with sheer weight of numbers which are forever increasing and with all being forced into what is inevitably a limited land space which cannot be enlarged, in some parts of the large urban conglomerations worldwide, that good reason is not now very far away. The massive predominance of private motoring over public passenger transport has brought with it conditions on roads in urban areas which have become well known and which can now be summarized:

- Traffic congestion has lengthened journey times and estimated arrival times are unreliable.
- Off-street parking becomes oversubscribed.
- On-street parking where permitted narrows the roadway available to through traffic and exacerbates what is already an unsatisfactory situation.
- Residential parking in side roads makes them at times virtually impassable.
- Loss of traffic has made public transport unviable which necessitates either subsidies, reduced facilities or higher fares with the latter two options driving away yet further traffic.
- Congestion makes bus services still less attractive.
- Resultant inferior public transport causes hardship to those who cannot drive or own a car, for example the young and the old.
- The private car with its average occupancy of only about 1½ people is extravagant in road space which is in short supply.
- Drivers unable to park become frustrated, park illegally, accept the penalty if caught and in the process create further congestion and aggravation.
- Parking concessions granted to the disabled create further congestion.
- With high density traffic and frustrated and inexperienced drivers, accidents involving personal injury or death are frequent occurrences which makes the road the most hazardous means of transport.
- With low-occupancy vehicles, fuel consumption per passenger mile is high which is an extravagant use of finite resources.

- High fuel consumption per mile causes unnecessarily high atmospheric pollution with its attendant health hazard and contribution to global warming.

Over the years numerous measures have been taken to alleviate the traffic problem, the most important of which are:

- Traffic management schemes involving revised traffic arrangements to make better use of the available road space. Examples are introduction of one-way streets, circulatory workings and banning right (or left) turns which cross the oncoming traffic flow.
- Traffic engineering schemes designed to give similar benefits to the traffic management schemes but more ambitious, and involving also new road works and maybe new connecting links.
- Construction of new roads and by-passes.
- Construction of new motorways.

However, all of these measures are but palliatives. They provide a temporary easement but the benefits are quickly absorbed by yet more vehicular traffic and the overall situation continues to deteriorate. And whatever the politicians or the motorists might like to think, land space is finite.

OPERATING REGIMES

Commercial freedom

Commercial freedom means no government intervention on the quantity issues; in other words, freedom from any statutory restrictions on the provision or amount of public transport services, their routes, frequencies, vehicle types or fares and charges. It means free competition. It does not (or should not) mean freedom from quality control, that is, regulations made in the interests of safety. This is the climate in which private as opposed to public enterprise can flourish in trade generally. Small businesses are able to compete with larger organizations in quality of service, levels of charges and ability to seek and satisfy demands. The theory is that the resultant facilities will be optimized by the action of free market forces and prices will be forced down.

It has been said that most public transport was developed on a commercial basis. Competitive conditions prevailed when two or more services of the same or different modes set themselves out to cater for similar traffics. Over the years customers became accustomed to the availability of choice, particularly with regard to the mode of conveyance, and this applies to the carriage of both passengers and goods. As the industry expanded and competition grew stronger, operating units became larger and more powerful. A process of amalgamation veered towards monopoly conditions, although the availability of choice between the modes generally remained and the public desire for such choice was still strong.

By custom and practice or personal preference without any specific reason, some people decline to travel by one mode but accept another. Some are discerning about safety and comfort although it is unfortunate that safety should be a consideration. In an ideal world no operators should run their services in a dangerous manner and, in any case, regulations should prevent them from so doing. Nevertheless, where such legislation is either non-existent or not properly enforced this concern is understandable. In some of the developing countries fatalities amongst bus passengers caused by badly maintained vehicles, speeding or overloaded on heavily congested or tortuous roads are frequent occurrences. Furthermore, human nature is such that individuals tend to have more confidence in one form of transport than another. The spectacular but horrific disasters involving roll-on roll-off ferries of recent years, large-scale fatalities on overloaded ferryboats in the developing world, the less serious but nevertheless unpleasant mishaps involving cruise liners and the oil tanker disasters have done nothing to instil public confidence in shipping, even though the chance of being involved in a similar incident in the future may be remote. Some travellers are still nervous of flying but are prepared to ride in cars even though casualties involving occupants of private cars are larger in comparison. There is not, therefore, a lot of rationale about modal selection on safety grounds but people do, nevertheless, have their preferences. Comfort is also a consideration which involves not only the quality and size of the seat and the smoothness of the ride but also the extent of crowding, the need for changing vehicles and the amount of walking and waiting at interchange points. The one transport requirement that nobody seems to care much about is the carriage of mail, perhaps because there is usually only one carrier of letters. Senders may not even know how they are conveyed, being interested only in the speed of delivery, and they will pay a surcharge for an allegedly faster service. Nevertheless, the postal service notwithstanding, it is clear that whatever the advantages and disadvantages of competition might be, there would seem to be a demand for the retention of modal choice.

Maximum service at minimum cost within a competitive environment is what commercial freedom is all about. If, however, the resultant facilities fall short of the ideal, compromise is not easy and it is at this stage that government intervention becomes necessary.

Co-ordination

Notwithstanding the fear of monopoly with possible public exploitation by a single supplier, history does not show that free competition in the provision of transport facilities is the ideal solution. The thinking that competition encourages enterprise and is a spur to improvement of quality and lowering of charges cannot be applied to transport in the way that it is to the manufacture of durable goods. Circumstances are not the same. On all pre-scheduled transport services which must run as advertised, the product is highly perishable. Unlike tangible goods, it is something that cannot be stored. Contrast this with, for example, the production of Christmas cards for peak sales in December. This activity can be spread over the entire year with stocks stored until they are required. Once seat miles have been produced (ie, the vehicle is making its journey) those seats must be sold on the spot. If they remain empty then their revenue is lost for ever. This does have repercussions on transport provision as any excess is wasted far beyond what might be the case in other industries.

It has just been said that local passenger services are suffering from abstraction of traffic by the private car and increases in costs result in part from the congestion that it causes. If, therefore, essential services are to be maintained at minimum expense, it becomes questionable whether resources should be allowed to be underutilized and wasted unnecessarily. It could be argued that wasteful duplication should be avoided if costs are to be kept as low as possible and hence fares maintained at a reasonable level. In other words, the services should be co-ordinated. To co-ordinate means to bring into order as parts of a whole or to put into harmony. In transport this can be achieved for the carriage of passengers and merchandise on a modal or inter-modal basis for a locally defined area or nationwide. Where, therefore, parallel competing services are provided they could instead be presented as a single entity with, say, the timings of the hitherto competing operators reasonably spaced with common fares and the acceptance of each other's tickets. The other important issue which again has just been mentioned is the now very topical feature that the movement of vehicles, vessels and aircraft pollutes the atmosphere. The supply of transport services, however crucial they might be, is, therefore, to varying extents, environmentally detrimental. Proper co-ordination will minimize any surplus provision.

By way of illustration, consider the cross-Channel services between England and France. Looking back into history, for many years the cross-Channel packet boats were owned and operated by the British and French Railways (both then in the public sector), supplemented on one route by the erstwhile private sector Townsend Thoresen Company. Commercially, the railway-owned boats were fully integrated with their shore-based operations and were an integral part of the 'railway' network. Alongside this, Townsend Thoresen was making great play of its private sector status, extolling the alleged virtues of its competitive enterprise.

But if that regime had continued through to today, then, with the advent of the tunnel, the railway authorities would now presumably be owning and controlling both the trains through the tunnel and the boats (apart from Townsend Thoresen). Under these circumstances it is suggested that the members of the British Railways Board, for example, might not be witnessing their trains and their boats competing with each other between Dover and Calais whilst spending (and losing) a lot of money in the process. However, with the two modes privatized and separated, that is the spectacle that emerged. Had there been no fragmentation, it is more likely that BR would have channelled its Dover–Calais traffic to the extent that it could be accommodated on to the trains and redeployed any surplus boats to other routes where they could have been used to better advantage. Some vessels might even have become redundant and sold or at least not replaced. But whilst that is a lesson in itself, it is not the end of the story.

Prior to the opening of the tunnel, privatization and acquisitions had led, as far as the British operators are concerned, to the presence of two competing shipping lines on the short sea routes from Dover: Stena and P&O. But then along came a third competitor, the railway, not with boats but with real trains and at that stage things became too hot for the free forces of competition. There came a plea from both of the shipping companies for Government permission to merge which would put them in a stronger position to remain in business. They needed to benefit from the greater economy in operation that a co-ordinated approach would bring. This is very different thinking to that of the erstwhile Townsend Thoresen, the one-time protagonist of free enterprise! And P&O is now on record[1] as saying that the combined venture (with Stena which was Government-approved only after considerable delay) is a prelude to a 'major cost cutting programme'.

Bearing in mind all the circumstances of this particular case, perhaps after all there is something to be said for co-ordination, even by private enterprise when the going gets tough.

The systems concept

Co-ordination could go a stage further with two or more operators merged into a single organization and all operations fully integrated under a common management, even to the extent of combining the modes and exploiting the best features of each (although, as has been seen, the elimination of choice of mode might not be popular). If this principle, conveniently designated the 'systems concept', prevailed then the services of road, rail, inland waterway, sea and air would not be planned in isolation but each would instead become part of one comprehensive system and the modes would become complementary to each other. In these circumstances, each would be used as appropriate for the particular job and for which it is most suited. Buses, for example, would where possible be used as feeders to rail rather than as a through service and merchandise would be conveyed by the most expeditious means (as post offices do today) which in theory would achieve maximum economy at the expense of loss of choice. It is likely also to involve large-scale ownership and control.

The systems concept which some may see as an ideal has never really been attained. Across the world there have been varying degrees of government intervention but, as has been observed, the desire for choice is strong, particularly between the modes.

Overview

Put simply, the main difference between the two systems described is that competition offers a haphazard and often bewildering collection of transport services. These are provided according to the law of supply and demand but probably at a competitive price whilst co-ordination produces a unified, comprehensive system, but this could restrict or deny the availability of choice of an operator or occasionally even a mode. The unified system is likely to offer a wider network without concentrating only on the busier and hence the more remunerative areas, but there would be less incentive for cheap fares. Both concepts, therefore, have their inherent dangers. If private, competitive enterprise had its free hand, then good facilities could be expected on the busy routes at popular periods but there would be areas and times of day when there might be no service at all. If a protected unified network was left to its own devices other than a requirement to provide an adequate and properly integrated service within its area of operations, then something nearer to a monopoly situation would prevail. This might blunt the spur of initiative and with the absence of direct competition there could be higher fares. The reactions and benefits or otherwise of these two strategies to the users is considered in Chapter 6. Either way there is a need for some controls. In a competitive situation any lack of provision might need to be filled, whilst in near monopoly conditions there would need to be a watchful eye on service provision and fares. More is said about this in Chapter 7.

In short, each system has its advantages and disadvantages and examples appear in this and in other chapters. No one system is likely to be satisfactory without some form of statutory requirement designed to ameliorate the worst of any adverse features. One outcome of all this is that although the political role can be and often is overplayed, some intervention is necessary. Certainly this is something that will need to be considered when planning for the future and it is, therefore, material for Chapter 9.

PLANNING

Prologue

When reference was made at the outset of this chapter to the environment, attention was drawn to the haphazard way in which older established towns have developed. This is partly because such controls as did exist were based on ill-conceived notions and partly because there were no properly considered overall plans. Remember the adage 'to fail to plan is to plan to fail' and many of the results of a past lack of or improper planning are still with us. Not only is the resultant land use structure sometimes environmentally unsatisfactory, it has in some cases also produced undesirable transport operating conditions. That is why railway lines and stations are sometimes in places inconveniently sited at a distance away from the towns that they purport to serve. Worldwide, when railways were built government authority was required, if only because they invariably needed compulsory land purchase and this put interested parties in a position to object. Influential landowners at the time often did not welcome and successfully resisted such intrusions on to their properties. In any case, it was argued, why should their sedate horse-drawn carriages have to give way to these evil iron monsters! Such controls as there were, therefore, were not always constructive.

Planning in this context falls into two parts. There is land use planning for which the provision of transport services should have an input and there is the planning of the actual transport facilities. Both are important subjects in their own right and beyond the realms of the basic principles, but general introductions are nevertheless appropriate.

Land use and landtake

Following on from what has just been said, the remedy for haphazard growth is organized planning of, among other things, the parallel development of land use, transport and related factors with a view to the achievement of optimum benefits from what are separate disciplines. The design and location of residential areas, for example, in relation to industrial estates, schools and commercial and shopping centres has a strong effect on the pattern and hence the cost of the provision of local passenger transport. The planning stage of new development is the time to incorporate space for a segregated light rail track or separate busways if either of those modes are contemplated. Unnecessary expense to bus operators has arisen in the past because of the unsatisfactory layout of built-up areas. With this reasoning it is good sense to allow transport to be at least one of the dictates of land use. The corporate planning process takes into account the requirements of not only the vehicular traffic but also of pedestrians and the environment. It is nevertheless imperative that adequate provision is made for penetration by public transport into places where the passengers want to go and not just to where it happens to be convenient to take them.

Organized planning and the resultant controls do meet with resistance in some quarters where it is seen as placing yet more restrictions on life generally. But in already densely populated areas with the numbers still growing but where the space is not, control is inevitable. Under these circumstances, no anti-public transport lobby must be allowed to predominate. Public consultation as discussed in Chapter 6 has relevance here.

A topical point on this issue is the provision of out-town shopping centres or hypermarkets. Shopping areas of this kind, which consist of one or more large supermarkets with adequate approach roads and ample (free) car parking facilities situated in what is otherwise green field territory, is something which tends to be frowned upon by some planning authorities and indeed by some national governments. Certainly they do rather split the resources of the retail trade and if a sufficient amount of patronage is lost by shops in the established commercial centres, it is feared that these areas would be liable to wither and die. There are reasons, therefore, why applications for development of this kind are not always welcome and chambers of trade in particular see detrimental repercussions. It is a concept that is popular in North America and has spread to Europe and now well beyond and into the developing countries. But a wish to preserve the character of the original town centres has caused many authorities to set their faces against these projects. Nevertheless, if the number of cars in the car parks is anything to go by, they are popular. They are highly convenient for those who wish to shop by car and that means a lot of people. Thoughts on their future is discussed in Chapter 9.

It has been said that transport should be one of the dictates of land use. Transport, being an essential service, must have adequate provision. On the other hand, transport is, to greater or lesser degrees, environmentally unfriendly, not only in the use of energy and its pollutant effects but through its landtake. Its infrastructure is visually intrusive on the environment. In modal terms, as far as landtake is concerned, air and sea transport are the most efficient for obvious reasons (even though the individual airports need a lot of land): the efficiency of the landtake assessed by the extent of occupancy of land by the relevant infrastructure expressed in persons or tons of freight over a given distance. After air and sea come the inland waterways and the railways. Last of all in terms of efficiency of landtake are the roads. According to UIC[2] (see Chapter 7) the capacity of a double track high speed railway line is equivalent to that of a road dual carriageway with four lanes in either direction.

Transport services

Transport networks need to be the subject of longer term planning. However, circumstances differ between continents and between countries and it is not surprising that thoughts and preferences also vary. Some areas (which are often the developing countries) are less conscious of environmental considerations, probably because being environmentally friendly generally involves spending more money. Also, and probably for the same reason, they often give less thought to the welfare of the individual. Neither do they burden themselves with the considerable concessions that are made to the old and the disabled as is increasingly being done in the West.

Other than the occasional high speed train networks and enlargement of airports, that part of transport planning which is uppermost in the public eye and which is now usually open to public consultation as referred to in Chapter 6 is the local urban systems. Selections, proposals and recommendations are made by the powers that be from the various modes which are considered in Chapter 2, unfortunately sometimes more for publicity and local aggrandizement rather than sound economic or practical reasons and all to a gullible audience attracted by glossy brochures. There is always the danger that environmental friendliness and disabled friendliness, which are the popular vote-catching excogitations of the day, are made the key issues, with the light rail systems, low-floor buses, mini-buses, cycle tracks and pedestrian walkways coming in for good measure as close runners-up. This is all very noble providing they do not overshadow the more mundane but nevertheless crucial matters such as overall cost, means of payment, value for money, service availability, maximum practicability, convenience and comfort for the majority and, above all, satisfactory movement of the masses. These important features are not always properly addressed. Certainly, the resultant provision must then be as environmentally friendly as possible. In places where traffic saturation point is rapidly approaching and roads are desperate for car restraints, a public transport plan will need to provide for even better facilities and likely passenger requirements are considered in Chapter 6. Examples of public transport plans follow.

Published planning proposals

Having discussed the basics of public transport planning, it is now opportune to look at a selection of the published transport planning proposals and strategies that have been prepared by various competent bodies. The presentations are basically to show the kind of things that a small sample of different authorities include in their strategies and objectives but the factual information that they contain is also of relevance and interest in this examination of the principles of transport.

European Commission
In 1993 the European Commission prepared a bulletin on the future development of a common transport policy.[3] This document constituted a global approach to the construction of a Community framework for sustainable mobility and considered numerous aspects including transport economics, objectives and then the various issues with possible solutions. All transport modes were considered. The salient features of this report are now briefly summarized with particular note being made of the specific information which is helpful in this study.

The declared fundamental objectives of the Commission's policy in respect of European transport towards which it intends to aim is a single market for transport services free of bureaucracy and quantitative restrictions (see Chapter 7) and ensuring free competition. Other major features are better financial performances and efficiency of the different transport undertakings with improved functioning and quality which includes safety, reliability and comfort along with measures to protect the environment. This is set against a background of increasing demand wherein it is stated that the demand for transport, be it for goods or passengers, generally runs broadly parallel to the growth in the gross domestic product (GDP). Between 1970 and 1990 the annual economic growth in the European Community (EC) averaged 2.6 per cent in real terms and for the same period the growth rates of transport services averaged 2.3 per cent for goods and 3.1 per cent for passengers. Indicative of the worldwide situation, it is noted that the increase of net disposable income and demographic changes have led to higher rates of car ownership and increased holiday and leisure time travelling. Nevertheless, the variations between the modes are markedly different, and in the movement of freight, road transport in this period has more than doubled in absolute terms and now accounts for about 70 per cent of all transport activity. On the passenger side, similar trends are indicated. Over this period, the bulk of the increase is due to the use of the private car which has doubled in absolute terms (which corresponds with the information contained in Figure 1.1) with its share of the market increased to nearly 80 per cent! The report goes on to reproduce data in respect of atmospheric pollution and other environmental intrusions which is illuminating. It is considered in Chapter 8.

Having set the scene, the bulletin sets out possible solutions to the predicted problems together with objectives and targets considered necessary to achieve a common transport policy with sustainable mobility for the Community as a whole. Environmental considerations have a strong influence on everything that is said.

A move beginning with the elimination of barriers and progressing towards an effective integrated system is anticipated and reference is made to the various modal features (see Chapter 2). Road has emerged as the dominant means of transport causing congestion which in some cases is approaching saturation of roads, it being alleged that prices do not reflect the full social cost. Railways are seen to have a potential to accommodate more traffic as the development of the high-speed network (see Chapter 7) will release more capacity on existing lines, whilst inland waterways, which are particularly efficient for bulk traffic have not as yet been widely exploited for other trades. The growth of air traffic (domestic and international) has severely strained some parts of the infrastructure: non-standardized air traffic management and control systems coupled with a lack of available runway and terminal capacities at some airports has resulted in delays which are likely to get worse. The problem is likely to be exacerbated by competitive pressures which could lead airlines to opt for higher frequencies with smaller aircraft. For maritime transport, the Commission considers that users concentrate on this mode where road transport, for physical reasons, is not a viable alternative and that coastal shipping should be further exploited. This, of course, brings into question the adequacy of the ports with the need to ensure adequate facilities and minimal delays.

Other than what has been said, much of the bulletin concentrates on welding the transport policies and services of the constituent countries of the EC into one common transport policy; this has a more political implication and is not strictly relevant to this study of the principles of transport as such. However, a further example of proposals emanating from the European Commission including what are suggested as good public transport planning principles is the Green Paper 'The Citizens' Network', also produced by the European Commission,[4] in this case in 1996.

In this document it is noted that car ownership within the European Union has increased from 232 per 1000 people in 1975 to 435 per thousand people in 1995 (again this corresponds with the information at Figure 1.1) and that private cars now account for 75 per cent of the total miles travelled. The social element of transport and its importance for people without access to cars is recognized, it being estimated that in Europe there are some 100 million elderly people, 80 million of whom are mobility impaired including some 50 million who are disabled. Half of the elderly people do not have access to private transport. Notwithstanding these figures, many more people in Europe do now have access to cars and the consequent ease of movement has encouraged urban sprawl which in turn has necessitated longer journeys between home and places of work and recreation. The environmental consequences are seen to be serious and in the Green Paper it was estimated that transport causes '62 per cent of carbon monoxide (CO), 50 per cent of nitrogenoxide (NO_x), 33 per cent of hydrocarbon and 17 per cent of carbon dioxide (CO_2) emissions'. The 1993 bulletin, however, recorded that transport was responsible for around 25 per cent of total CO_2 emissions. Increasing road capacity would not reduce pollution but would exacerbate the already inefficient landtake of the private car. The construction of more roads in an attempt to contain the growing demand is not, therefore, proposed as a solution.

A move to more public transport is seen as a way forward and (to quote from the bulletin) even at an occupancy rate of only 50 per cent, energy consumption per passenger mile of bus and regional rail services still makes it about five times lower than for private cars. This, of course, is quite apart from the landtake issue where, on a per passenger basis, buses are claimed to require only 5 per cent of the road space required by cars and fuel consumption per passenger is clearly substantially less. The main parameters towards a 'citizens' network' are regarded as accessibility, affordability, security, convenience and environmental impact. Guidelines are suggested which both encourage the use of public transport and discourage the use of cars, and emphasis is given to the need for co-ordination between the modes whilst retaining the element of choice.

Good practices for passenger transport planning as an aid for developing the citizens' network are reviewed and discussed. Interestingly, it submits that for shorter journeys of up to about 3km (which is approaching 2 miles), walking offers a viable alternative with pedestrian segregation being desirable. Bicycles, which are considered to be a suitable alternative to the car for journeys of up to about 5 miles, should be integrated into the system which calls for better provision for cyclists in terms of both infrastructure and shared road space. Other features reviewed are:

- taxis; which tend to be underutilized although they can in some circumstances be cost-effective, particularly if shared;
- private cars; for which a development of transport telematic technologies within the framework of a broader integrated road transport environment would assist. Also, measures to increase vehicle occupancy rates are considered appropriate;
- public passenger transport; the improvements for which should cover the key areas of vehicles and rolling stock, system integration, information provision, quality of service, increased convenience, planning priority and land use planning within an integrated policy approach. Special reference is made to midi and mini buses which are able to serve once inaccessible or newly pedestrianized areas, light rail and guided busways.

The highlights of the proposals are:

- systems integration, to bring in all of the modes as a part of a common network;
- timetables, with everything connecting with everything else;
- tickets, allowing for through bookings from here to there to any-where;
- multimodal terminals, being a crucial part of the integration process;
- information about everything to be freely available, including during the course of a journey;
- door-to-door service, to be as nearly as possible as good as the convenience of a car;
- planning, priority to public transport, to put public transport in a more competitive position compared with the car;
- reserved lanes for public transport, to provide for buses as well as trams;
- priority for buses at traffic signals, with a description of the systems;
- restricting car access, including banning cars from central zones;
- land-use planning, aiming to ensure that businesses and services with a high potential of public transport utilization by staff and visitors are sited where they can be made easily accessible by public transport.

These are the perceived ideals and best practice examples which the European Commission has compiled as a basis for its citizens' network. The bulletin then goes on to considerations for disseminating know-how and setting targets, aligning research and development priorities with user needs, making Community policy instruments effective, moderniz-ing the regulatory framework and improving standards.

King County Metro (Seattle, Washington, USA)
By way of introduction it can be said that Seattle, King County, is a good example of an area which has already planned to encourage the use of public transport and much has already been achieved. It gives free travel within the Central Zone and a tunnel has been constructed across down-town under Third Avenue which buses and trolleybuses use to avoid the traffic congestion in the town centre. Hence the advantage of the dual-powered vehicles described in Chapter 2 which allow suitably equipped buses to use the overhead wires in the tunnel. Its entire active fleet is lift-equipped and thereby accessible by the disabled (which, if the operator is in receipt of money from public funds, US law requires it to be) and all vehicles are being fitted with bicycle racks. Fares outside the centre are roughly comparable with or perhaps below Western world standards gen-erally. With free fares in the centre (and reduced rates for the elderly and disabled elsewhere) only some 25 per cent of the total costs is actually recovered from passenger receipts which indicates that Seattle is prepared

to give considerable financial support to its public transport. The fact that the scheduled vehicle requirement between the peak and off-peak is at a ratio of 2.2 to 1 shows that a considerable amount of costly works and business commuter traffic has been attracted to public transport.

King County Metro has produced a six year Transit Development Plan which covers the period 1996 to 2001[5]. Services implemented under this plan provide still better quality services to more people more often and ensure response to changes in growth patterns, travel behaviour and community needs which are expected to continue through to the next century. Prior to 1996 the system could generally have been characterized as providing a large proportion of radial services from the centre to relatively few major destinations and strongly orientated towards a commuter market. The modified service pattern builds on to the most successful parts of the network, whilst at the same time being expanded to serve new areas. A 'multi-centred' system focused on a series of transit 'hubs' where convenient connections can be made to more destinations (on the same principle as some national airline networks) has been implemented during the past three years which provides a greater area of accessibility.

London Transport (England)
London Transport (LT) is the owner and through its subsidiary the operator of the local rail rapid transit system (the 'Underground'). It also authorizes the routes, timetables and fares of the buses, the services of which are put out to tender. It has a statutory obligation to produce a statement of strategy every three years and as one further example, the 1997 statement[6] will now be reviewed.

The principal declared objectives of LT over the three-year period and longer are summarized as follows:
Shorter term objectives (1–3 years)

- improve on the Underground and work with bus operators to improve on the buses the quality and reliability of services and increase off-peak provision;
- continue to improve integration through better ticketing, interchanges and information;
- complete the Jubilee Line extension (an Underground line) and the Croydon 'Tramlink' (an LRT line);
- aim to achieve the Government's targets for continuing service and financial improvements.

Additional longer term objectives (beyond three years)

- move towards achieving LT's vision of Modern Transport for London to maintain London's position as a world-class city;

- achieve adequate and consistent funding needed to recover from investment backlog and deliver 'Modern Transport for London';
- complete the priority network schemes for buses and seek re-allocation of road space;
- contribute to improving London's environment by promoting use of public transport and minimizing LT's negative environmental impact;
- increase revenues by improving services and continue to improve operating efficiency thereby progressing towards nominal financial self-sufficiency;
- expand the network with new and extended rail lines and intermediate mode schemes.

The 'vision of Modern Transport' consists, by and large, of what is listed in the other objectives and which are contained and enlarged upon in other LT planning documents. It is noted that as fiscal policy and the allocation of road space are the preserves of central and local Government, the various authorities must work together if the measures proposed by LT can win an increased share of the travel market for public transport and thereby achieve reduced congestion and an improved environment.

Much of the strategy centres around funding (or the lack of it – although there has been an allocation of capital since the report was prepared). At the time it was written it was noted that necessary major improvements to the Underground would be possible only through the Private Finance Initiative (a principle discussed further in Chapter 4) or other public/private partnerships and where a strong case can be demonstrated. LT would develop projects in this way where there was good value for money. LT states that the funds that it generates are dependent on the demand for its services, the level of fares and the cost of provision. LT expects the overall demand for travel to increase by some 15 per cent by 2011 with a 20 per cent increase in rail commuting. Unless, therefore, adequate steps are taken now it is anticipated that even in the short term there will be significant problems of congestion on the Underground. LT intends to continue its policy of increasing fares slightly above inflation with the proceeds invested in more and better services, as it considers it is reasonable that passengers meet some (but not all) of the improvement costs.

The two other main features of the strategy are traffic congestion and the environment. There is a declared wish for more car restraints with parking restrictions and discouragement to use cars to be effected through the land use planning process. In environmental terms, the strategy involves (like Sydney – see Chapter 5) use of the lowest emission diesel engine and ultra-low sulphur diesel fuel. Also, the policy is to fit older buses with oxidizing catalysts which enables them to match the new low emission standards.

One further example of a corporate plan with management objectives is contained under the sub-heading 'Objectives' in Chapter 5.

Characteristics and Components

INTRODUCTION

It was seen in Chapter 1 that transport comprises various modes which together constitute a system. Collectively, transport meets the demand for the movement of people and goods (in economic terms it satisfies the utility of place) but the nature and circumstances of these demands differ widely. Each mode has its own characteristics which determine suitability for the conveyance of different commodities or people over different journeys. Practicality is of course a major factor, followed by cost, speed and in the case of passengers, comfort. These are the features which motivate choice. Then there are the characteristics of the different types of traffic that must be carried and the services that are provided.

MODAL CHARACTERISTICS

Road

The fundamental characteristic of road transport is its flexibility and physical ability to provide a door-to-door service. With minor exceptions such as railway private sidings, waterside premises and helicopter landing pads, it is the only means that can give such a service, hence travel by any other mode invariably requires carriage by road at the extremities. Apart from its value as a means of trunk haul the road is, therefore, also essential for feeder purposes for rail, water and air. As the use of the public highway is open to all it means that the road is a shared way with many different users and as many different types of traffic. Certainly the road is not unique in that its use is shared. However, the fundamental difference between the road and the other forms of way, and it is this that underlines the importance of the sharing principle, is that the road is eminently suitable for private transport as was considered in Chapter 1. Because of its sheer convenience the car is a highly popular alternative to public transport and the number now on the roads is phenomenal. Furthermore, they are mostly in the care of non-professional drivers, many of whom (unlike the other modes) although 'qualified' are inexperienced and unskilled! This does have considerable operational repercussions.

There are two reasons why the road is particularly suitable for private and personal vehicle ownership; one being that, in many countries, the capital cost of a car is within reach of the majority and the other that it is not too difficult to obtain a licence to drive. It is these properties coupled with the ability to provide a convenient door-to-door facility that has endeared the private car to so many people. Motoring has become a way of life, the results of which are considered in Chapter 8.

The same problems have also arisen with the carriage of goods, again because of the ease of obtaining a vehicle, the relatively low operational costs and the greater convenience. In the case of goods, however, there are more professional drivers.

A particular feature of the public highway is that it is public in the most liberal interpretation. Other than on the special motorways where access is prohibited for anything other than driving a vehicle, normal highways and any adjacent footpaths are available for everybody. The road is freely used by all types of traffic; fast and slow, large and small, passenger and goods, bicycles and pedestrians. In built-up areas, the road, although intended as an artery along which to travel, is sometimes also used as a car park, and this is not just for short stay whilst shopping or conducting business; it is often for permanent residential parking! Thousands of cars have the public highway as their only permanent base. Worse than that, in the East such as in the Indian sub-continent, roads in towns are also used by people as the fancy takes them for commercial and domestic purposes. Space does not sometimes exist anywhere else. Food stalls, car repairs, haircuts, garbage dumps, etc all compete for a few square feet of the roadway along with the needs of general living such as cooking facilities and charpoys (wooden framed string beds). All of this activity together with children, beggars, cows, dogs, goats and chicken that see the road as their rightful habitat has to be circumnavigated with the aid of accelerators, brakes, horns and word of mouth by a motley collection of smoke emitting large and small buses, large and small lorries, cars and three-wheeled motor scooters (small taxis) plus tongas (horse-drawn two-wheeled 'taxis'), camel carts, bullock carts, bicycles and pedestrians. This reduces what might otherwise be a wide thoroughfare into something narrow and congested. The resultant environment may not be healthy but at least it is colourful and exciting.

Because of the profound multi-user aspect of the public highway, vehicle size and speed limitations are more severe than is the case with the other modes. Given comparable operating conditions, therefore, neither speed nor capacity is a feature of road transport; both rail (particularly high-speed trains) and air are superior in these two respects. The specialized motorways, however, have done much to redress the balance and, depending on circumstances, for some journeys road might well be faster than rail. Nevertheless, it has been said that road transport can provide a door-to-door facility whereas movement by the other modes requires transfer at terminals. When taking this into account, then, depending on the length of the trunk haul, the complete journey by road could again be quicker. The advantage of road transport, therefore, lies in convenience (door to door without involving transshipment or feeder services) and cost (low overheads). Even then, some road hauliers are likely to transfer goods between local collection and delivery and the main trunk haul.

A further point, however, is that whilst to the user transport by road is often the preferred choice, to life in general it is becoming socially harmful. As the average occupancy of a car is only some 1½ persons, it is extravagant in its use of road space and (in terms of passenger miles) also in the amount of fuel that it uses. That in turn means more atmospheric pollution and, as petrol and diesel are oil-based, a greater consumption of oil which in due course is expected to become scarce. In environmental terms, the car is unfriendly and more will be said of this in Chapter 8.

It is often heard said, particularly from the rail lobby, that road transport does not properly meet the full costs of its way and that it has an unfair advantage over rail which has to bear its entire track costs. What are very heavy (sole user) track expenses in the case of a railway are allegedly paralleled by just the cost of an annual vehicle licence in the case of road transport with the roads then being provided and maintained at public expense. The merits or otherwise of this allegation will, of course, vary country by country according to the different forms and levels of taxation. It is not practical, therefore, to portray the situation worldwide. As an example, however, the British Road Federation (BRF), which is the UK affiliated organization to the International Road Federation referred to in Chapter 7, makes the point that the only logical line of approach for this purpose is to incorporate all of the taxation which is incurred by road users and which is specific to them. Accordingly, it has produced figures in respect of Great Britain which take all relevant taxation into account.

BRF's relevant figures[1] in respect of Great Britain for the year 1996/97 are as follows:

	£m	£m
Specific road user taxes		
Yield from		
Fuel duty	17,400	
Vehicle excise duty	4,300	
(Vehicle licence)		
Total	21,700	21,700
General road user taxes		
VAT on vehicles	3,600	
VAT on fuel	3,000	
Total	6,600	6,600
Grand total		28,300
Total road expenditure for the same period		5,450

Source: BRF extracted from TSGB

Whilst it is true that in this British example the revenue just from vehicle licensing (£4,300 million) fell short of total road expenditure (£5,450 million), other forms of taxation specific to road users took their contribution way above what was actually spent on roads. In fact, in this instance it was nearly £23 million more. On the basis of these figures, therefore, less than 20 per cent of the amount which was actually paid by British road users in 1996/97 in the various forms of relevant taxes was actually spent on roads. Furthermore, the UK Government is committed to increasing fuel duties in the future by at least 5 per cent per annum above the rate of inflation.

Certainly this is an example of only one country but some years ago the European Federation for Transport and the Environment undertook a similar exercise in respect of 11 European countries for the year 1989/90 which, although now dated, did produce a similar trend.

There is certainly nothing anywhere here to support the contention that road transport does not meet its own track costs.

So much for the characteristics of road transport. Although popular, the road has its limitations. With vehicles under independent and personal control, congestion has become rampant and journey times unreliable and there are repercussions on the environment. Public transport does sometimes use the public roads for specialized purposes such as buses in the form of guided busways referred to in Chapter 3.

Trolleybuses and trams are also cases in point and the characteristics of these modes will be examined.

Examples of a selection of the many different types of vehicles that use the public highway are included in the Plates but a brief reference can be made here to the fuel that is used. Whilst it is basically petrol for the light vehicles and diesel for those that are heavier, concern is now being expressed regarding their pollutant properties. Different types of fuel are, therefore, now being introduced, particularly on buses, including, for example, liquid petroleum gas (LPG) and compressed natural gas (CNG). These alternatives and others are discussed further in Chapter 8.

Another feature which affects bus body design is a possible segregation of passengers. Occasionally in some of the developing countries separate accommodation for superior class passengers may be provided even on local buses (or trams). Furthermore, some of the Muslim-predominant countries may segregate male and female passengers. Local urban buses in Pakistan, for example, have separate compartments for this purpose which only goes to show how local customs and practices have their repercussions on vehicle design.

Trolleybus

The trolleybus is a hybrid, being an attempt to combine the flexibility of the motor bus with the electric propulsion of the tram which in many cases it superseded. The result is a vehicle similar in appearance to a bus but equipped with poles to collect electric current from overhead wires. It is therefore still tied to its tracks but not as rigidly. It is steered in the conventional manner within the limits of its trolley booms and it could also participate in guided bus lanes. But temporary diversions cannot be undertaken at short notice and special turning arrangements are needed. Interestingly, to effect a turn at Hampton Court (England) at the time of a tram to trolleybus conversion scheme in 1931, a roundabout was constructed at a hitherto T-junction. This heralded a new type of road layout at busy intersections and is now in widespread use. However, for emergencies most trolleybuses are equipped with batteries which enable them to move albeit slowly for short distances off the wires.

The trolleybus is silent, it has good acceleration and is particularly suitable in areas where there are heavy gradients (San Francisco and Seattle in the USA are examples); unlike a street tramway it is able to set down and pick up at the kerbside and take its place in the normal traffic flow. But like a tram it does not emit poisonous exhaust along the highway and, depending on how electricity is generated, it does not necessarily have to rely on oil. But like all electrically propelled systems, high capital costs are involved as the power must be transmitted from the generating stations and fed through sub-stations into a specially equipped track, being the overhead wires. Not every route within an urban network is, therefore, likely to be so equipped. But if not, it means two separate systems – part trolleybus and part conventional bus – with the attendant inherent disbenefits. The trend has therefore been to abandon trolleybuses. But because the electric motor is exhaust-free and quiet it is more environmentally friendly than the diesel bus. There are, therefore, reasons why the trolleybus should not be discarded lightly; there is further discussion on this issue in Chapter 8.

Some trolleybus systems do still flourish worldwide, particularly in parts of Europe, China, Russia and to a lesser extent North America. In Great Britain the once sizeable networks have all disappeared in the interests of cost, flexibility and standardization. But if the environment is to have priority over costs then there might be reason to think again about the abandonment of trolleybuses. Their overhead wires are to some people visually intrusive but it is submitted that in downtown areas which are so often cluttered with high-rise buildings, garish commercial advertising, road overpasses, elevated roads and railways, etc, trolleybus wires should not really be so objectionable (even if they are noticed).

Recent developments have sought to maximize the advantages and minimize the disadvantages of the trolleybus by introducing a dual-powered vehicle. To overcome the cost of equipping all routes with trolley wires, including the sections which are more lightly used, the dual-mode vehicle has been fitted with both an electric motor and a diesel engine. This enables it to travel either under power or independently. It is a more expensive vehicle both to buy and to maintain and could have a small capacity penalty. But if only the busy parts in the central zones of the main towns are electrified with the more lightly trafficked sections in the outer areas worked by diesel, then much of the cost of the overhead wiring is saved and the inner cities are freed of some of the exhaust.

Examples where dual-mode vehicles have been introduced are at Seattle (USA) by Metro Transit (whose development plan was referred to in Chapter 1) and at Lille in France.

Seattle has both diesel-powered buses and electrically powered trolleybuses of varying sizes as well as the dual-powered vehicles – the latter constitute almost 20 per cent of its total fleet of 1270 vehicles. In common with USA practice, all are single-deck.

The operating costs per mile of each type as quoted by Metro Transit vary between US $1.03 and 1.57 depending on size and type, in the case of the diesel bus, $1.68 in the case of the trolleybus and $2.35 for the dual mode. These figures include fuel (or power) and a proportion of tyres and maintenance except that in the case of the trolleybus, they do not include maintenance of the electrical infrastructure.

Tram

The street tramway with its tracks set in the public highway was a satisfactory means of urban transport at the turn of the nineteenth century and it remained in widespread use until the 1930s. Being a railed vehicle, however, it is, like the railway, tied to its tracks. As general vehicular traffic increased, one school of thought was that the tram could no longer take its place in normal traffic flows and that it was materially aggravating congestion. Furthermore, as the tracks were generally laid in the centre of the road, other traffic blocked its path which only added weight to this line of thinking. Partly for this and partly for economic reasons, tramway networks have now been abandoned in many areas around the world. In some countries conventional street tramways have entirely disappeared, whilst in others there may be only a few or even just individual systems left. With the mode having become so scarce, where trams do remain, certain isolated networks come easily to mind. For example, in India think of Calcutta, in China of Hong Kong, in Japan of Hiroshima, in Australia of Melbourne, in Canada of Toronto, in Egypt of Cairo and Alexandria and in England of Blackpool. Then there are the countries of northern and eastern Europe and across Russia and Ukraine where many of the larger towns too numerous to mention still have flourishing tramway networks, all of which do an excellent job in providing local transport.

Why, then, has the abandonment of the apparently discredited tram not been universal? Why has the sporadic retention of trams been done in such a haphazard manner? Why is it that the surviving systems are able to make a valuable contribution to the transport needs of the local communities and why is it that there is currently much talk about the resurrection of trams with some newly developed systems in the form of light rail transit already working?

It was during the second half of the nineteenth century that horse buses first took to the streets in any numbers. However, their large wooden iron-rimmed wheels on cobblestone roads produced a resistence to rolling which meant that a 26-seater body, although inadequate in both physical and economic terms, was the maximum that two horses could pull. Something better was therefore needed. It was known that a horse could pull a much heavier load in a barge on a canal as there is less resistance on water than on a road; it was this principle that had already justified the construction of canals. By the same thinking, a horse could pull a heavier vehicle if it had steel wheels running on smooth steel rails than it could if running directly on the uneven surface of the highway. Accordingly, right from the beginning it was a matter of capacity. Then came electric power as was referred to in Chapter 1; systems were electrified and the route mileage grew, reaching a peak around the 1930s or, in other words, until the omnipresent traffic thrombosis prompted thoughts of replacement by trolleybuses or buses. But circumstances were not the same everywhere. Private transport did not develop as early or as rapidly in Russia and eastern Europe, for example, as it did in the West which meant that such traffic as there was flowed more freely and there was no clamour to remove street tramways. Even in the more heavily trafficked places a street tramway could be made to work if it was accompanied by proper traffic management measures with a degree of segregation. In some cities that segregation was already built and in those parts of Europe that were devastated by war it was possible and convenient to make such provision at the time of reconstruction. In some cities, however, to segregate tram tracks from the existing roads would have involved substantial demolition which could not be tolerated. As the track, the electrical equipment and the trams became time-expired and due for renewal (at considerable cost), there was need for a decision on whether the trams should be replaced by some other, probably much cheaper, mode.

The tram is expensive because the tramway company, like the railway company, has to bear the costs of providing and maintaining its own specialized way. But the railway, unlike the tramway, has the advantage of not having other traffic running across its tracks. When tram lines are set in the public highway, the tramway company is in the invidious position of bearing the full costs of its way whilst at the same time, as has been said, sharing it with all other road vehicles which can freely impinge on its tracks. Unless the track is segregated, therefore, the trams are exposed to the same degree of congestion as is the rest of the traffic – and many systems were abandoned to make room for cars! But although the tram is expensive and inflexible to the extent that it cannot easily be altered to meet changing demands or even to pass another vehicle which obstructs its special way, it does have its advantages. Like the trolleybus, it does not necessarily rely on oil for its source of power; it does not emit exhaust along the road and it has a high carrying capacity. With its high cost of construction, routes are more likely to be forced on to fewer and more concentrated radial 'spokes' to fan out towards the extremeties and which, some will say (including the writer), is a good planning principle. Everybody knows where to find the tram route; people adjust their travelling habits accordingly and gravitate on to what is a simplified network, accepting perhaps a longer walk in return for a more frequent service. Even its inability to divert is not always a bad thing as, unlike buses, there can be no demand for trams to move across to another road. The permanent nature of the way discourages local politicians, town planners and the like from preparing grandiose schemes for the pedestrianization of city centres with public transport sent round the back. Such schemes have come and it has been easy to push out the buses but the trams have stayed where they are (look at Hanover in Germany and Melbourne in Australia, for example) and by continuing to penetrate those places where the buses are no longer allowed to reach but where their passengers still want to go, they are doing a good job. At the same time they are more environmentally friendly in those otherwise traffic-free areas. Too often the bus has been a victim of its own flexibility.

There is yet one further feature in support of the tram. The point was made in the Road sub-section above that in some of the developing countries urban streets have multifarious unofficial uses with the result that traffic has to thread its tortuous way through with difficulty and delay. But for psychological reasons, people do not seem as willing to impinge on visible tracks (be they railway or tramway) to set up their stalls or place their charpoys. They might get within a yard (metre) or so of the rail but no nearer. This means that the tram can often get through easier than can the bus. This situation was illustrated in Karachi (Pakistan) where at one time there was a local tram system (it was unique in that the trams were diesel-powered). When the trams were abandoned and the tracks went into disuse or were covered over, the opportunity came for the local populace to find its own extraneous uses for the little extra space that had become available and the traffic did not then flow quite as easily.

A feature of some significance is that trams can be double-ended, that is, they can be driven from either end and are able to do something that no other road vehicle can; with the provision of a simple crossover in the track or running on to a single line at the terminus they can reverse direction in the centre of the road. However, not all trams are double-ended as there are penalties with this type of construction. If they are not, then special turning circles must be provided in the same way as they are in the case of trolleybuses.

What has been dealt with so far is essentially a street tramway. It was said, however, that there is now a development of this mode in the form of light rail transit. This will now be considered.

Light rail transit (LRT)

The distinction between a tramway and a railway is blurred. A vehicle running on tracks laid in the centre of the road and providing a local service is clearly a tram. A segregated high-speed line with express trains covering long distances is clearly a railway. But there are numerous stages in between. The special reservation and on to the road here and there can still be regarded as a tramway and the suburban commuter network of a main line system is still a railway. However, there is a trend now to introduce new railway type track or to develop disused or lightly used local railway lines for use by tram type vehicles supplemented by new sections which encroach on to roads just to serve important traffic objectives such as town centres, etc (it was said in Chapter 1 that the original railway stations were not always conveniently sited). This type of system is therefore something more than a tramway, the dictionary definition of which is 'a passenger car running on rails laid in a public road'. Then comes the entirely segregated system, be it on the surface, elevated or even underground but again with a lighter type of vehicle than that used on a conventional railway. That also should not be regarded as a tram. There are plans afoot worldwide to bring in more systems of this kind but the popular conception of this development, being that 'the trams are coming back' is not seen by the writer as being particularly accurate.

Light rapid transit bridges the gap between the buses and heavy 'metro'. The lighter vehicles can negotiate sharper curves and steeper inclines than can the heavy metro and its ability to deviate into popular traffic objectives enables it to match the convenience of the bus. It is these systems that are categorized 'LRT'.

Monorail

An alternative light urban rail system but with less application is the monorail. There are isolated examples where monorails are used as part of an urban transport system but the application is very limited. Its main use is as an internal distributor in shopping malls, pleasure grounds and the like. The monorail is not a system that is likely to make any substantial contribution to urban transport and only if a demand arises with a traffic potential insufficient to justify the cost of a full-scale underground line but too great to be absorbed by surface transport within an area where the land topography is suitable might a case for a monorail be proved.

Rack railway

A rack railway is a specialized type of railway that enables purpose-built locomotives to climb hills that could not be negotiated by normal adhesion. A steel cogged rail is laid between the running rails to which pinions on the train engage and so control movement. This type of railway is able to negotiate steep gradients and is used in mountainous regions. The Pilatus Railway in Switzerland, for example, which was opened in 1889 has a maximum gradient of 48 per cent. This particular line is electrified but the principle is not confined to any one source of power.

Heavy rapid transit

Heavy rapid transit is the mode associated with mass transit in urban areas. Its construction, if underground, involves one of the most complex and costly building operations of any mode but the resultant systems, where they exist, make a crucial contribution to the smooth functioning of a large densely populated conglomeration.

The term 'rapid transit' is applied in transport parlance to heavily used urban railway systems, parts or preferably all of which are segregated from all other services, including other services of the same mode. They are often designated 'Metro' or 'Subway' or 'Underground'. In this context, the importance is that the paths of trains on different services do not conflict because if they do, then any delay or irregularity on one line can be transmitted through to another.

As populations increase and urban conglomerations get bigger, congestion on the roads becomes a problem and part of the answer invariably lies in the construction of new underground railways. However, an underground line also has to be affordable and with the high capital costs involved the mode is unlikely to be viable in its own right. Revenue may cover day-to-day running costs but it is unlikely that capital costs involved during the course of construction will ever be recovered. But there are other benefits and the subject is referred to again under the subheading 'Cost-benefit studies' in Chapter 4.

Capacities of the urban modes

At this juncture and before proceeding further with the remaining modes which cover the longer distance journeys, it is opportune to give some idea of the potential capacities of the different modes which are a part of urban transport. It is of course unrealistic by any rule-of-thumb method to determine absolute maximum capacities in terms of passengers per hour on any one type of system. Whatever the declared maximum there is usually room for one more. Another passenger could crush into a railway coach; another bus could find its place on the public highway. What follows, therefore, constitutes generalized observations more than scientific fact based on formulas compiled to interpret specific situations. The thoughts are presented only to give a measure of perspective and must be treated accordingly.

Buses and trolleybuses

Although technically different, buses and trolleybuses have similar characteristics insofar as their passenger-carrying ability is concerned. Remember that buses share the public highway with all other vehicles, passengers and goods. The amount of traffic that can be moved by a conventional bus service is not, therefore, determined solely by the qualities of the vehicle itself but more by the highway on which it runs. Highway capacities are governed by such things as width of road and number of traffic lanes, general speed of traffic and obstructions to progress such as conflicting vehicle movements, traffic lights, pedestrian crossings, parked vehicles and the like. If a busy shopping street is compared with a motorway the differences become apparent.

The capacity that a bus service can provide is nevertheless also governed by the size of the vehicles and their frequency. As far as the vehicle element is concerned, the capacity of a conventional-type bus is finite, that is, it is limited to the number of seats plus standees (hopefully not very many). For a standard double-deck this could be about 80 but very large double-deck or articulated crush-loading vehicles can carry many more than that. To ascertain what is the maximum capacity that such a service could provide it is necessary to know how many vehicles can actually use the road in question and that is the figure that is elastic and depends on circumstances. If, for example, with standard double-decks there was a 30-second headway or 120 per hour (and that is a lot of buses) then nearly 10,000 people could be moved in that hour providing they were evenly spread as any peak surges within that hour could not be accommodated. Apart from highway capacities, the provision of stops and waiting areas could also become a problem. A service of this level would be better with a segregated bus lane (or even a guided bus lane). If this in turn provided an exclusive uninterrupted flow – ie, grade separated junctions, etc – and if such a facility was provided for the full length of the route then it gets into the light rapid transit class. With extra large crush loaded articulated vehicles the figure would be still higher but it would offer a very uncomfortable way to travel. In any case, these figures represent the extreme, they are not likely to be attained in practice and the traffic is not likely to be evenly spread. There is usually a peak within a peak and the crucial factor is more what can be moved within the maximum quarter hour.

To be realistic, in any one hour some capacity is invariably wasted and a sustained 30-second headway without some form of delay and consequent bunching is super-optimistic. A practicable figure for hourly capacities attainable, certainly on the normal public highway, might be more in the region of, say, 5000 for standard double-decks and 7000 for large double-deck or articulated buses (or about 1250 for midibuses!). This is substantially less than the theoretical maximum quoted but here there might be a little more room to accommodate short additional bursts of traffic.

Tram
Although the tram is at a disadvantage compared with the bus in terms of cost, it can have a greater capacity. Whilst the articulated tram can carry

about 160 people, albeit in discomfort, there would seem to be no reason why two double-deck trams, each with 100 seats or more to eliminate standees, could not be coupled together and run in pairs with one driver, particularly if the track was largely segregated. At this stage, the tram becomes more akin to LRT. Indeed, this is precisely what happened on the long-since-closed Swansea to Mumbles Railway in what is now West Glamorganshire in Wales. The vehicles were double-deck, they sometimes ran in pairs, the track was segregated; it was then classified as a light railway and today it would have been categorized as LRT.

A double unit of this type running every 30 seconds would give a theoretical hourly capacity of about 25,000 or more people. However, like the buses, a regular 30-second service, even on a partially segregated system, is unlikely to be attained. With such fine precision, a slightly extended wait at a stop whilst passengers are alighting and boarding or even at traffic lights would result in a block back from which it would be difficult to recover and would lead to yet wider gaps in the service. But based on a 60-second headway the theoretical capacity could still be 12,000 or more per hour. If the line was completely segregated with, say, four cars coupled together, then even the original figure would become more realistic with the line now coming firmly into the LRT class.

Heavy rapid transit

Heavy rapid transit, completely segregated and grade separated represents the ultimate in mass transit. The movement of trains is governed by a signalling control system which may often now be automated and it is the arrangement of this system that governs the length of trains. But generally, 7–10 car units can travel at speed between stations say a mile (1.6km) apart or even less with down to about a 90-second interval between them. A system such as this can provide realistically for about 30,000 passengers per hour but a feature of heavy rail rapid transit is that it does have an overload capacity and hence the ability to meet sporadic heavy surges of traffic. Some 15,000 people in a maximum quarter hour (the equivalent of 60,000 per hour) becomes physically possible and is certainly not-unknown. Again, however, even on a 90-second headway, there is little margin left to counteract delays and, with so many people travelling under cramped conditions of this kind, something is always likely to happen. A slightly extended wait at a station will delay the train at the rear, more people will have assembled at the next station to make alighting and boarding even more time-consuming and then comes the inevitable gaps in service and bunching. That is why the suggested overload of 15,000 people in 15 minutes should not be accepted as a norm for an extended period.

One further point of note is that when it comes to traffic of this order it is rarely possible to plan for every passenger to have a seat for the whole of the journey. It has been said that this is the ideal but under these circumstances, something less has to be accepted and it is not quite as arduous to stand in a railed vehicle as it is in a bus. In terms of passenger calculations, therefore, figures are based not on passenger seats but on passenger places which takes into account floor space and an agreed number of standing passengers per square yard (square metre). As the number of people allowed for (and hence the standard of travel) tends to be more liberal in the Western world than it is in the developing countries (ie, a lesser number of people per square yard of car floor) this is yet another reason why it is not possible to give precise figures.

Overview on capacities

Once again let it be stressed that these figures are intended only to give some sort of order of magnitude. The picture has been painted with a broad brush. But in any long-term planning exercise, it may suggest what sort of a system to look for once the travel predictions and likely future demands have been determined.

It is of interest that STIB (Brussels Transport), which is concerned about the inordinate amount of road space required by private cars, has also prepared capacity figures but has expressed them in a different way.[2] It is stated that its Metro line under rue de la Loi in the centre of Brussels (Belgium) carries 20,000 people per hour in each direction in a tunnel 25 feet (7.5m) wide containing one double track with trains at 3-minute intervals with 1000 passengers per train. Based on this example, it estimates that to carry this traffic by the other modes, the highway (in this instance, rue de la Loi) would have to be enlarged as follows:

For cars: A 300 feet (90m) broad roadway to give 15 lanes in each direction to carry 1000 cars in each lane with each car carrying, on average, 1.33 people. (Note the photograph at Plate 1 which depicts a 3–4 lane highway in each direction and compare this with the 15 lanes in each direction that would be necessary in the centre of Brussels!)

For buses: A 60 feet (18m) broad roadway to give 3 lanes in each direction to carry buses at 45-second intervals in each lane with 80 passengers per bus.

For trams: A 40 feet (12m) roadway containing two double tracks to accommodate trams at 90-second intervals along each track with 250 passengers per tram.

Quite clearly the first alternative would be a sheer impracticability and the trams would be hard pressed.

As one further illustration, the trams in London (England) in their heyday in the 1930s had a terminal loop working round the Victoria Embankment between Westminster and Blackfriars which had one double track. Along this double track there was a total of 14 different routes which ran in each direction and which fanned out to different parts of south London. During the maximum peak hour a total of 148 trams were scheduled to pass along those tracks each way. The trams had 74 seats and were allowed 9 standing and most of them did carry that number of passengers. But it would have taken only 68 per tram to have accounted for 10,000 passengers in the hour with everybody having a seat.

Finally, an interesting but not particularly useful piece of information is in respect of cable-cars. A cable-car link is to be constructed 250 feet (75m) above the River Thames from Docklands to the Millennium Dome at Greenwich in London. Each gondola will carry 15 passengers and the declared capacity of the line is 2500 people per hour.

Railway

By its very nature a railway must have its own specialized way, terminals (stations), signalling and all the necessary apparatus to ensure the safe passage of trains. In consequence the provision of the way requires a substantial capital outlay and as the equipment must be kept in good working order there is a heavy maintenance bill. Overheads, therefore, represent a high proportion of total costs. On the other hand, the railway is a private way in the sense that it can be used only by its specialized vehicles. Although, in contrast to road transport, the railway is naturally inflexible and it alone must bear the entire cost of its way, the reward is that the trains, unlike the buses, are not required to share their way with all and sundry. This does give it some operating advantages. The railway operator – or track authority if it is separate – has complete control of all movement along its lines and it is thereby in a position to enforce a strict code of operational discipline. Subject to an appropriate signalling system, high-capacity trains can work a frequent service at fast speeds unencumbered by a multitude of other less disciplined users which is in sharp contrast to the roads, as has already been noted. The larger units also enable the provision of more elaborate passenger amenities on the trains such as restaurant cars, sleeping cars, luggage accommodation and toilet facilities. A modern railway is therefore eminently suitable for medium and long distance traffic with a good trunk haul where its speed and capacity potentials can be properly exploited. But the expensive stock and equipment must be kept fully employed if the service is to be anything approaching viability. It follows that both density and distance are the required features of railway traffic, hence the policy of closing branch lines and intermediate stations in rural areas and concentrating on freight in bulk and large numbers of people for which the railway is competitive in speed and comfort.

Physically if not profitably the railway is also very suitable for coping with heavy surges of traffic for special events and also for the concentrated morning and evening peak business movements in the conurbations. Local urban railway networks cater for this type of traffic. Beyond that are often the specialized and segregated heavy rapid transit systems (the features of the specialized infrastructure are discussed in Chapter 3), although suburban networks of conventional main line systems do also have a strong input to this type of operation. By catering for heavy commuter movements the railway fulfils an important social function, as a parallel road system could not accommodate such traffic and there would be serious or even chaotic congestion. As populations increase and urban conglomerations get bigger, congestion is a problem that can only worsen and part of the answer could lie in the construction of new underground railways. In view of the high capital cost of construction, such systems would not be profitable in the accountancy sense but, depending on local circumstances, they might be the most desirable solution. Reasons other than commercial viability could therefore influence railway development.

To summarize, in economic terms passenger traffic on the railway is ideally channelled on to a relatively few main trunk routes serving the larger towns and fed by buses. Outside the built-up areas and again in economic terms, local intermediate traffic seldom justifies the costs involved. Even if a service is maintained with the cheapest possible vehicle – a diesel railbus – the track must still be maintained at a cost far beyond the cost of the shared portion of the road borne by buses, to say nothing of the signals, stations, etc. As far as freight is concerned, it is the type and quantity of goods that spell profit or loss for the railway. Also, like passenger traffic, intermediate goods traffic is expensive because of the need to provide full facilities at wayside stations to meet small and irregular consignments. Again, therefore, merchandise must be channelled along main lines and between points where traffic is consistently heavy. Most profitable are regular complete train loads with automated or minimum handling of commodities which lend themselves to mechanized loading and unloading.

Most of what has been said so far relates to the railway in economic terms. In other words, it is selecting its optimum uses, physically and financially, and leaving the rest to go some other way. But it was seen when discussing road transport that the time is approaching when that 'some other way' may no longer constitute an acceptable alternative. Roads are in parts approaching saturation and it is environmentally damaging to construct more. In other words, the railway profit and loss accounts cannot now be considered in isolation. The wider implications of social needs and the environment must become an input and hence a part of a broader spectrum and this may well influence the size and scale of future railway networks.

Pipeline

The pipeline is unique in that the way is also the carrying unit thereby producing a transformation from batch-handling to continuous flow. It is inflexible as once it is laid its position cannot easily be altered and this calls for a sustained demand between fixed points. It is not, therefore, suitable where movement patterns are liable to change. Furthermore, its use is to some extent specific as it is necessary to be selective in grouping materials for conveyance through the same tube. Whilst it is technically possible to separate one product from the next with virtually no mixing in transit and there is no reason why a delivery of white spirit should not be followed by a consignment of, say, paraffin, the same pipe would not then be used for, perhaps, a delivery of beer.

The pipeline does have advantages compared with the more orthodox forms of transport. As the way is also the carrying unit no packing is necessary and hence no returning of empty containers. Nothing moves except the commodity being transported. The problem of a return load does not arise and traffic remains unaffected by road congestion, transshipment points or difficult terrain. The elimination of a separate carrying unit and the absence of handling makes the pipeline light on manpower. The safety record of pipelines is likely to be superior to the alternatives which is an advantage when conveying hazardous fluids such as ethylene and petroleum. Also, in terms of capacity, the pipeline is likely to be superior to either road or rail. It has, therefore, much to offer in return for its capital cost and general inflexibility.

The cost of movement by pipeline is also different to that of the other modes. The cost of batch road or sea transport does not vary materially with the quantity transported once the largest possible vehicle or vessel is kept fully employed. Costs are then at their lowest and any traffic variation means only a larger or smaller number of carrying units. The cost of pipeline transport drops rapidly with the quantity moved until saturation of the line's capacity is reached and when no more traffic can be accepted without another pipeline. The high initial capital cost is then spread over the maximum volume of traffic. Owing to its static nature, operating costs are relatively low and are largely made up of fixed elements such as labour, maintenance, wayleaves, etc. Then, as is the case with the other modes, the optimum operating costs are obtained when the pipeline operates continually at full capacity. There is nothing strange about that. What is more significant is that once the pipeline is used to its maximum capacity, the direct operating costs compare favourably with other forms of inland transport. Pipeline economics become particularly favourable in harsh environments and remote locations such as cold, desert and jungle regions where there is no existing infrastructure or where alternative modes are unsuitable.

The oil industry has developed the use of pipelines for the bulk conveyance of oil and natural gas. In fact, gas in any quantity can be moved only by pipe. For the conveyance of oil, however, the pipeline is an alternative to batch transport either by waterborne tankers on the sea and inland waterways or by rail or road tankers. Although as has been said the economics of the pipeline can compare favourably with road or rail transport, even with the biggest pipeline capacities, the cost need not come down to the level of ocean-going tankers for the same length of haul. The pipeline is, however, competitive with tankers if its route is considerably shorter, if the sea journey involves exceptional charges such as heavy canal dues or if a special harbour has to be constructed for the exclusive use of this particular traffic.

These reasons explain why the principal use of pipelines is to connect oilfields with refining or shipping centres. Pipelines are, however, able to convey suitable semi-solids and solids either suspended in liquid or pneumatically. Also, they may be reserved for the exclusive use of the products of one manufacturer or, alternatively, the owner may undertake to convey traffic for other people. In other words, the pipeline may be used for either public or private purposes depending largely on whether the owners are manufacturers who require their own private facilities for distribution or whether their business is that of a public carrier.

Sea

The fundamental characteristic of sea transport is that the way is natural subject to the provision of navigational aids and free to use, but the terminals are specialized. It was said in Chapter 1 that the first means of transport was on water and it was those early shipping activities that forged the pattern of our present trading routes. Today, a large part of seaborne traffic is international, much of it being carried over long distances. For natural reasons, international shipping gives a service that often cannot be accomplished either easily or at all by any other means except air. Vessels have a high capacity and apart from the influence of the elements can provide a high standard of comfort for passengers. But sea transport is traditionally slow and the longer distance passenger business other than cruising has now been lost to air. Bulk carriers, container ships and tankers now form a major part of the world's mercantile marine which carries some 90 per cent of world trade. Passenger traffic has, however, been maintained on short sea routes where journeys would otherwise be difficult or physically impossible except by air. Examples are the UK to Europe; Germany and Denmark to Sweden; across the Mediterranean Sea; between the numerous islands which constitute the Philippines and to the offshore islands of Japan. For this type of traffic, roll-on roll-off ferry boats enable road vehicles of all sizes (passenger as well as goods) to drive directly on and off the vessel.

For the movement of merchandise the great capacity of seagoing vessels makes the mode eminently suitable for the conveyance of bulk cargoes and the ability of ships to carry good payloads offsets their high cost of construction. Shipping rates can therefore be maintained at reasonably attractive levels and sea transport remains strong in the movement of merchandise, particularly where the journey cannot be made by any other form of transport except air. But it is an environmentally friendly mode of conveyance and in the future governments may well have to look for a revival of coastal shipping as an alternative to road transport.

It is pertinent to note in this context that the (British) Royal Commission on Environmental Pollution[3] is on record as saying that transport by water is the preferred mode in environmental terms even to the extent of offering financial support from public funds. Scotland was cited as an example where it is less expensive to support shipping than it would be to upgrade parallel minor roads to make them acceptable for heavy haulage vehicles. A case was also seen to improve port facilities and even to support the shipping companies themselves in order to facilitate the transfer of freight away from the roads and on to the sea in the interests of the environment. Note also the thoughts of the EU mentioned in Chapter 1. To illustrate the value of coastal shipping, the Royal Commission quoted Japan where the utilization of this mode of transport is fully exploited. The policy has been motivated by a lack of adequate roads in certain areas of Japan coupled with a maximum lorry weight of 25 tonnes. It is significant that neither of these factors had retarded economic growth in Japan which shows that for an island nation or where otherwise geographically appropriate, coastal shipping has much to commend it and not only in environmental terms.

A feature of the transport of general merchandise is the time taken to load and unload. A ship is a particularly large unit and its high capacity magnifies the time spent in turn-round at terminals. Remember also the safe arrival concept, which means that goods must be delivered without loss, damage or delay. It is at the points of transshipment where these three possibilities are most likely to occur. This is the vulnerable point and a through facility that eliminates physical transfers between different modes can only be beneficial. The economic advantages of any reductions are, therefore, great. The roll-on roll-off ferry boat just mentioned with its resultant creation of what is in effect a through transport facility between road and sea (and can also be between the railway and sea) has been beneficial in this respect. This is particularly so when it is used (as it usually is) on the short sea routes where loading and unloading times hitherto represented a high proportion of total time including that spent at sea. The various forms of mechanization in the ports have also made a strong contribution to this end. However, this is also one of the reasons why the container has developed to the extent that it has.

Containers offer a direct facility between the major inland ports of origin and destination and are another form of through transport. Like the roll-on roll-off ferry, they also eliminate the need for transshipment of individual consignments at ports. They are unique in that the same container is equally acceptable on road and rail as well as at sea and the principle could therefore have been referred to with equal relevance under either of those modes. However, for convenience the concept is mentioned here as there is an extensive use of specialized container vessels. Because of their flexibility, containers are able to take maximum advantage of each of the modes according to the geography of the journey as it is the container and not the carrying unit that follows the required path. Although the goods that are carried are likely to be of assorted shapes and sizes and may consist of a series of miscellaneous consignments, the container itself is a standardized size worldwide. It is an easily stackable unit regardless of the degree of loadability of its contents. The vehicles and vessels that carry containers are suitably constructed for the purpose. As far as transshipment is concerned, although there are similar advantages to those of the roll-on roll-off ferry, the vessels are less complex. They do not need the controversial entry/exit doors at either end which are the weakest points of the boat and have been responsible for disasters in recent times. Again like the ro-ro ferry, specialized terminals are necessary. But for a container port only a large area of open land is required for stacking purposes with gantry cranes which enable rapid loading and unloading.

Containers are an excellent example of the systems concept to which reference was made in Chapter 1. They are also used for the conveyance of freight by air but as in this case they are not of the same standard dimensions, operations cannot be integrated with the other modes.

It was said at the outset that a feature of water transport is speed – or the lack of it. Boats are not fast. This adverse characteristic has, however, been partially overcome by the hovercraft, the hydrofoil and the twin-hull catamaran which are able to travel faster.

The hovercraft is amphibious which means that it can follow more direct routes regardless of depth of water (and hence tides) or, indeed, without any water at all. Its basis of construction is described at Plate 48. It can travel over any surface which includes mud flats, swamps, sandbanks, snow, level ground and hardstanding. It can come in from the sea and skim on to a beach or hard pad located well ashore for terminal purposes. It does not need a dry dock for maintenance and can park safely on its pad when not at sea. It is much faster than a ship, being capable of speeds of up to about 75 mph (about 65 knots or 120 kph). As an example of its versatile application, the hovercraft is useful in the far eastern parts of Russia along shallow rivers that are otherwise unnavigable due to ice for many months of the year. Its terminal requirements are simple as all that is required is an area of hardstanding for passengers to board and alight.

The hydrofoil which, unlike the hovercraft, can travel only on water, is used with particular advantage as a ferry boat across relatively calm waters to offshore islands and (more relevant to the next section) on lakes and along navigable rivers. The hydrofoil is equipped with foils (or 'wings') and it is these that create the lift. It 'flies' in water much as an aircraft flies in the air. The foils are outstretched once it is in open water but when entering or leaving port they are enclosed and the vessel then functions as a normal boat.

Another alternative to achieve higher speeds is the twin-hull catamaran vessel which again makes less contact with the water and therefore has less drag.

Inland waterway

As is the case with all water-borne transport, movement by inland waterway is not fast. For this reason it is more suitable for the conveyance of lower classes of goods of a non-perishable nature and for standing stocks that are subject to continual replenishment and for which there is no particular urgency. Coal, petroleum, grain, timber, chemicals, iron and steel lend themselves to this form of transport. Inland waterways are specially suitable for:

- traffic routed through ports connected with a waterway system, particularly where overside delivery from ship to barge or vice versa takes place;
- traffic that can be carried in bulk from point to point in barge loads;
- traffic conveyed to and from waterside premises or for storage in canalside warehouses;
- trunk haul to waterheads such as an inland port with subsequent delivery by road.

The viability of an inland waterway depends very much on its width and depth and hence the size and capacity of the vessels that use it. Whilst narrow canals which were constructed before the advent of road and rail transport now have no commercial transport value, the wide water arteries such as the River Rhine in Europe and the St Lawrence Seaway in North America are of strategic international importance. The latter opens up the Great Lakes to ocean-going shipping and the port at Duluth (USA) on Lake Superior is over 1000 miles (1610km) as the crow flies from the nearest point on the coast (at New York).

Although reference has so far been confined to the movement of merchandise, the carriage of passengers also has a limited place in this mode of transport. Waterbuses ply as part of an urban transport system where circumstances are appropriate as they are, for example, in Venice (Italy), Kerala (India) and Bangkok (Thailand). Some of the larger rivers such as the Rhine (Europe), the Nile (Egypt) and the Yangtze (China) are also popular for holiday cruising.

Air

The outstanding characteristic of air transport is speed. Not only is the aircraft itself very fast (apart from any political considerations regarding the use of air space), it is also to follow a direct route unhindered by the physical barriers experienced by surface transport. Its terminals, however, are extravagant in land take and can seldom be placed in the immediate vicinity of large towns. It follows that there must often be lengthy journeys by road or rail between town centres and airports. This together with terminal formalities is time-consuming and the portion of air travel within any one journey must be sufficient to make the overall time competitive with travel by surface transport throughout. For this reason, unless there are special circumstances or the use of specialized aircraft such as helicopters which land at places where conventional aircraft cannot, air transport seldom offers any advantage for distances of less than, say, 200 miles (320km). Air transport is therefore appropriate for the long haul (intercontinental and interglobal) but it could also be competitive for shorter journeys across country that is difficult for surface transport, physically or perhaps even politically as, for example, from Bangkok (Thailand) to Ho Chi Minh City (Vietnam).

In general characteristics there is a similarity between air transport and shipping which has just been considered. The reasons are:

- International routes predominate in both cases and both are of strategic international importance.
- The way is natural in each case and requires no capital investment.
- Both require an extensive system of navigating aids.
- Both require elaborate and specialized terminals with adequate access and which the operators do not generally own.

• Both have a large charter market.

But there the similarity ends. They are at the opposite ends of the scale in terms of speed. The air caters predominantly for passengers (but not entirely and air cargo is a fast-growing market) whilst traffic by sea is predominantly freight (but not entirely and cruising is a fast-growing market). In terms of capacity of the individual vessels and aircraft, it is of course shipping that has the very large units. But the difference is not quite as stark as might appear at first sight. Because of the speed differential, an aircraft can make many trips between any two points in the time that a boat is making one which materially boosts the comparative capacity of an air service.

TRAFFIC CHARACTERISTICS

Preamble

The fundamental characteristic of the traffic that is carried is, of course, whether it is passenger or goods; so much so that some carriers, notably in road transport, specialize in one or the other. Not only the vehicles but also the operating techniques are different and even in undertakings that carry both such as a railway or airline, the administration is at least partially separated. Traffics within each of these two broad divisions, however, also have their own different characteristics.

Passenger traffic

People use transport for a variety of reasons. They travel between home and their place of work or school and they may travel during the course of their work. This is considered to be essential traffic. Then there is the optional traffic which consists of those who are not travelling in connection with their employment but are doing so for personal reasons such as shopping or visiting friends, pursuing leisure activities, visiting places of interest or attending sporting events. The demand for optional travel is elastic which means that it does not have to travel at set times on set days and it is more likely to respond to the price and quality of the service. People in this category have less real need to travel and it is easier for them to find alternatives. They could shop locally and they could find other pastimes, even to the extent of staying at home. Over the past 30 years or so, television has encouraged them to do just that as an alternative to making short excursions. The transport requirements of the business community are much less elastic and can better be described as inelastic. Those travelling for their own private purposes at their own expense will be more interested in bargain fares and, in view of the elasticity of their demand, will wait for and respond to special offers which are made for off-peak travel. Those travelling on business (and on an expense account) might see comfort and convenience as their prime considerations and the cheaper fares are unlikely to be available at the times that they require anyway. In planning the facilities and the fares, therefore, it is necessary to identify the class of traffic at which they are to be aimed. The fares aspect is further considered in Chapter 6.

Distance is another characteristic. Works and schools traffic is often local and the journeys relatively short. Office staff sometimes travel further and the same might be said of special shopping expeditions. Otherwise, travelling in the course of business, for recreation, or for some special purpose could amount to anything from a local trip to something embracing the length of the country or the circumference of the earth.

It is unfortunate from the point of view of transport economics but nevertheless inevitable that the demand for travel is not easily spread and there are periods of acute traffic concentration. As far as the daily peaks are concerned, business traffic is likely to create heavy unidirectional flows, be it people on buses or cars on roads, with schools traffic conforming to no specific pattern but making everywhere busy. As far as public transport is concerned, to meet short peak demands with resources that are otherwise underutilized is uneconomic and costly. Private transport, in the form of private cars, offers relief in this respect but to reduce what has now become chronic vehicular congestion with all of the physical problems and environmentally damaging effects that that entails, the cost of putting this traffic back on to public transport is a price worth paying. The cost-benefit studies as discussed in Chapter 4 elaborate on this point. Other types of traffic have a more seasonal aspect such as holiday travel which concentrates at popular times, particularly during school holidays. Clearly more thought needs to be given to getting people out of their private cars and brought back on to the buses and trains. Suggestions to this end are contained in Chapter 9.

Passenger demands and expectations are assessed in advance which has repercussions on the types of service provided and which are discussed below. Unlike freight, passengers are standard units which in this respect simplifies vehicle design. Subject only to length (and sometimes class) of travel, all they expect is a seat and a comfortable ride which they do not always get. Although the expected quality of the ride may vary according to distance it must nevertheless be stressed that if public transport is to have any success at all in enticing people out of cars, that seat must be comfortable, even for short trips. However, this is something more for Chapter 6. The point being made here is that the nature of the traffic is such that for passenger vehicles there can be a considerable degree of standardization which is a very different situation to that in the case of vehicles designed for the carriage of goods. Passengers, unlike goods, also load and unload themselves which means that the terminal arrangements are also very different.

A special segment of passenger traffic is tourism which transport has made possible and which has brought in its wake the development of the hotel industry and other local amenities. Tourism, of which more will be said later in this chapter, stimulates local economies and it is something on which many of the smaller nations now rely.

Goods traffic

The characteristics of goods traffic are quite different to those on the passenger side. Unlike passenger movements, the demand is more sporadic. There are of course exceptions, such as contracts for bulk hauls, but miscellaneous consignments are dealt with on an ad hoc basis and this has marked repercussions on the economics of the transport service.

Goods may be in a solid, liquid or gaseous state, they may be inert or volatile, durable, fragile or perishable, high or low value, wet or dry or even dead or alive. They have an infinite variety of shapes, sizes, densities and weights. Much of the traffic requires specialized vehicles or vessels with similarly specialized handling methods and equipment and some goods cannot be mixed with others. With goods that are not so exacting in their requirements, general purpose vehicles can be used but any awkward shapes may demand excessive space in relation to weight. This brings into focus the term 'loadability'. Traffic that has good loadability has the property of being able to accommodate itself within and make maximum use of the space available within the vehicle, vessel or aircraft in which it is being carried or, in other words, it gives a good space/weight ratio. Coal, grain and sand, for example, have an excellent loadability characteristic.

Reference has already been made in this chapter to the use of containers and there are two aspects of loadability in this context. Containers are 'stuffed' with miscellaneous goods which will have varying degrees of loadability. But the container itself is of a standard size and shape which, as a unit to be conveyed, has perfect loadability even though what it contains may not have. This feature has particular relevance in the shipping industry where many containers may be conveniently stacked in a single ship although perhaps all carrying what could otherwise be awkward loads.

There is one further feature in relation to goods traffic to which attention must be drawn. Over the years the general conception has been that it is the passenger services that require speed of travel and which must be afforded priority of movement. The carriage of freight might be slower but as a general principle that was accepted. For that reason it was considered (and still is considered) that goods are suitable for conveyance by sea and inland waterway. But in part that is changing. Certainly, passenger traffic still requires priority of speed but now, also, so does some of the freight. At one time manufacturers and suppliers kept stocks of raw material, etc in hand for processing, sale or whatever. This, however, is not in the best interests of their economic working as it represents idle capital tied up and unproductive. It is more efficient and, even after allowing for increased transport costs, it is more economic to receive stocks precisely at the time that they are required. In other words, the goods need to be delivered 'just in time' (JIT) for immediate processing, onward disposal, etc. But the 'just in time' traffic (as it is now designated) is a somewhat new concept for goods transport. Speed and time of delivery become crucial and transport companies, both road and air, have geared themselves to this demand. Similarly, the demand for perishable foodstuffs out of season or not obtainable locally is growing which again requires speedy transit. For example, strawberries from Spain or lobsters from the USA to Great Britain are better without several days at sea and the transport industry is rising to the occasion both on the roads and in the air. It is unfortunate that the JIT principle is less environmentally friendly as more deliveries need to be made (probably with smaller vehicles) which are not necessarily filled to capacity and more fuel is thereby used.

SERVICE CHARACTERISTICS

Passenger and goods services

When considering the characteristics of a service, the fundamental distinction between what is provided for passengers and what is provided for goods is again present. Compared with the traffic classifications, the distinction becomes a little blurred when, as is often the case, vehicles, vessels and aircraft designed primarily for passengers also carry some freight and, conversely, when those designed for the conveyance of goods do sometimes carry a few passengers. There is, nevertheless, usually a basic purpose for running the scheduled services, be it for passengers or goods, even though it is not necessarily an exclusive use. Country bus services to a small extent and passenger trains to a greater extent carry unaccompanied parcels and mail. In thinly populated areas, particularly in the developing countries, what are basically freight trains might also have a passenger coach attached. Passenger ships and aircraft are able to carry a limited amount of cargo according to their configuration and instances of the converse can be found. Aeroflot (Russia) and Eva Air (Taiwan) offer examples where some of their passenger aircraft are adapted to accommodate more than the normal amount of freight. In rural areas, vehicles used for the conveyance of mail may also be available for passengers and cargo ships may carry up to 12 passengers without a special passenger certificate. Nevertheless, the distinction between passenger and freight services remains.

Scheduled and charter services

Another basic characteristic is that between a scheduled service and one that is provided on a contract basis. A scheduled service is planned and advertised in advance and will operate regardless of demand at the time. In long-haul sea transport, pre-scheduled shipping services (albeit now almost non-existent on the passenger side) are known as liner services and the term is also extended to other modes. Whilst scheduled services must adhere to their published timetables, vehicles (or vessels or aircraft) working under a contract (or 'charter' or 'private hire') are used in accordance with the wishes of the hirer (although depending on local laws the hirer might sell the seats separately). Freight of a general nature and which is not perishable can sometimes be diverted or delayed in a manner that would be unacceptable to passengers which means that even pre-advertised freight facilities are not necessarily operated in the same rigid manner as are passenger services. With goods being moved by prior arrangement, facilities can be suitably adjusted to conform to the differing day-to-day requirements as they emerge, which makes it easier to maximize payloads and minimize waste.

In the case of a contract, charter, etc, an inclusive charge is levied to cover both the direct and, at discretion, also the indirect costs of the complete operation regardless of the number of people or the load that is carried. Responsibility for solvency is in this case with the hirer and not the operator who is assured of his return. This type of operation is the basis of package holidays (discussed below) although the travel trade does now also use the scheduled airlines with block bookings.

Excursions, tours and packages

An advertised excursion or tour is something which really lies somewhere between a charter and a normal service. It is more like a private hire but organized by the operator or by a third party who sells separate tickets to each passenger. If sufficient patronage is not forthcoming a tour programme can be altered at short notice which can involve cancellations and transfer of passengers. It is a step towards better load factors as it does to some extent avoid the situation which arises on a normal pre-scheduled service where once an empty seat has passed a point it is wasted.

One of the early beginnings of the 'package tour' business was in 1841 when a local man by the name of Thomas Cook chartered a train from the then Midland Counties Railway Company to take a party from Leicester to attend a temperance fête in Loughborough in England for an inclusive sum and made a separate charge for each participant. The resultant price of a ticket worked out much less than the normal fare and the principle of the cheap excursion was born. The name of that enterprising gentleman has survived and today it is well known in the travel business worldwide. The arrangement was beneficial to the railway as it meant filling seats and utilizing equipment at times when traffic would otherwise have been low, such as on Sundays when the business community did not travel. The principle gradually expanded and special fares were introduced, geared to attract passengers at off-peak times to special events, places of recreation and to the sun and sea. But the distances travelled were relatively short and seldom crossed international borders. It was more than a hundred years after Cook's memorable journey that air travel for the masses became a reality. By now, aircraft were chartered and the travel trade was organizing cheap flights to even better sun and seas. But it was not just the flight alone that came to be organized. Complete holidays were assembled with one comprehensive price to cover not only the flight but also board and lodging, transfer arrangements and even insurance, all bundled together in a complete package. And so the 'package holiday' was born which has come to embrace all modes of travel.

With intricate pre-booking arrangements involving both transport and hotel facilities, including the use of specially chartered aircraft, tour companies are able to maximize the use of all resources and hopefully reduce to almost negligible amounts any empty and hence wasted seats and hotel beds. Costs are thereby spread over the maximum number of people which means that the charge to the individual is kept attractively low.

It is of course the popularity of tourism that has so dramatically expanded the demand for this type of travel which has enabled 'the average man' to see the world. At the same time it has stimulated local economies but it has also brought with it serious environmental problems. This issue is considered in Chapter 8.

Quality of service

Frequency, regularity, speed and comfort are the characteristics that passengers tend to use to measure the qualities of a particular service.

For a local journey in an urban area, an acceptably frequent bus service is likely to be one that runs at least, say, every 12 minutes. Passengers using this type of service are not likely to be timetable-conscious and this should give them an average wait of 6 minutes. For a rail journey of 100 miles, two journeys per hour could be regarded as frequent and passengers in this case would most likely be prepared to adjust their habits in accordance with the advertised times. For a flight from London to Australia, a regular daily service could be considered acceptable. Regularity is generally considered in terms of frequency, ie, the time that elapses between successive vehicles or, in other words, the 'headway'. But a regular headway that is exactly repetitive in each hour is really more than just a regular service, it is a regular 'clock face' service. Odd bus journeys may run for works or school purposes, a shipping line might run a vessel per week or per month on a liner service, but these are all nevertheless regular services even though the actual headways might not be repetitive.

Speed is something that again can be looked at in different ways. It was seen when the modal characteristics were discussed that some forms of transport have a greater potential for speed than have others. But some services may be regarded fast and others slow regardless of the mode and in contradiction of the generally accepted modal characteristics. This could arise, for example, where a change of vehicle is required by one mode of transport but not on another. If a transfer is necessary for a journey by train but not by bus, then in terms of overall journey time it might be faster by bus even though the maximum speed attained at any one point en route was greater on the train.

Comfort is something which can be attributed only in part to the size, style and softness of the seat (assuming that the passenger always gets a seat). There are other features. They include the conditions under which one has to wait (such as the kerbside or a railway station), whether a change is involved and if so the standard of the interchange facilities, the smoothness of the ride and whether one does indeed get a seat or if the conditions are standee with crush loading. All of these characteristics contribute to what is the conceived comfort of the service.

All of this is coming round to the point that, in judging qualities, no one service is likely to be considered in isolation. The users wish to move either themselves or their goods between two places. They are concerned with overall movement from the beginning to the end of the journey. Each individual service may constitute only one part of that journey; to the operator(s) several services may be involved but to the user it is one journey. In other words, users are interested in the assortment of services provided. Cheap fares marketed through the free forces of competition have already been noted. The availability of choice was discussed in Chapter 1 as also was the systems concept. These fundamental concepts tend to be pulling in opposite directions and it is the consumer issues that will be further considered in Chapter 6.

THE COMPONENT PARTS

A study of the different types and designs of vehicles, vessels and aircraft involves much detail and all are widely different. Like the basic infrastructure, they also can be dealt with in only a generalized manner and in any case it makes no sense to group them together in a single study. However, the illustrations with their captions make a useful introduction to the subject and all that remains to be done here is to consider the principles and the common threads.

The fundamental distinction between the carriage of passengers and the conveyance of goods was noted when the characteristics were discussed. Although there are designs to accommodate a mixed traffic, it is usual for all vehicles, vessels or aircraft to be arranged internally for one or the other to predominate. The few exceptions are found in countries or areas where traffic is sparse. As has been noted, it is then sometimes economically justified to accommodate both passengers and goods in one combined unit and accept the resultant compromise of a timetable that is endeavouring to cater for the different requirements of the two traffics. However, even with the mixed train which was mentioned, the individual units are adapted solely for one or the other.

Another feature is the size and capacity of each unit. Really without exception the general rule is that maximum economy means maximum capacity and bigger and better has long been the order of the day. That is, of course, subject to there being sufficient traffic available to utilize that large capacity. There is nothing economic about carrying a lot of empty seats around. Neither is it good policy to reduce a service to an unattractive level for no better reason than an endeavour to concentrate traffic on to a lesser number of vehicles and thereby justify the use of the larger units. But that notwithstanding, double-deck buses have developed from an early utilization of the roof to accommodate more passengers, and road haulage vans have grown to 44-tonne trucks as well as the 'artics' (articulated vehicles). Also, longer trains (with double decks on some suburban lines) on the railways, strings of barges pulled or pushed along inland waterways, mammoth ocean-going supertankers and 'jumbo jets' (Boeing 747s) in the air all give credence to this trend.

Then there are the different requirements within both the passenger and goods classifications; and this applies to all of the modes. For passengers it is a question of comfort or capacity and this will depend on the length of journey and the class of travel, assuming that there is to be more than one class. A discussion on the provision of different classes is contained in Chapter 6. In the case of urban road transport, it is for decision whether passengers are to be crush-loaded on to a single-deck bus or comfortably seated on a double-deck. Also in urban areas the degree of comfort to be afforded regardless of distance or class is for consideration. Much will depend on the type of area served and the standard of conduct of the users whether comfortable soft seats or hard and uncomfortable but more vandal-proof fibreglass seats are fitted on the local buses and trains. Again, Chapter 6 elaborates on passenger preferences.

The point has been made that goods present an infinite variety of shapes, forms and sizes and vehicles and vessels are often specialized for the conveyance of particular commodities.

THREE

Infrastructure

PREAMBLE

There is a relationship between the characteristics of the various modes and the required infrastructure together with the different types of vehicles (or vessels or aircraft) that are used. The characteristics were discussed in Chapter 2. This chapter follows on from there and examines in general terms the required infrastructure of each mode.

The different types of way are roads (which includes certain fixed track systems which can also use the roads), railways, inland waterways, sea and air, the infrastructure requirements for the latter two being in the form of terminals. Each of these different modes will now be examined.

ROAD

Public highway

A road is something that may well have grown out of a one-time footpath. When people wished to use it more often and for longer distances it became a bridleway, being a way along which horses may be ridden. Then in due course it became a properly surfaced but initially narrow lane and thence progressed through various stages of development into what in some cases have become major traffic arteries. The paths or bridleways often took devious routes in order to skirt private property and this led to road systems that were in parts tortuous with sharp bends and severe gradients and when faced with a significant increase in vehicular traffic were no longer satisfactory. The result has been a continuous process of widening and realignment involving encroachments on to land either side.

According to traffic conditions and amount of use, a road may be narrow to the extent that oncoming vehicles cannot pass or wide enough to permit a centre reservation dividing the directional flows with two or more traffic lanes either side. The narrow single-track road might have special passing bays provided at intervals as can be found, for example, in the north of Scotland. Alternatively, where any attempt at all is made to provide a sealed road the verge of a metalled single lane could be flanked by a tract of hard level ground on which vehicles can encroach in the (rare) event of meeting as, for example, in the Australian outback. Another further cost-cutting arrangement and what can sometimes be found in the rural areas of some of the African countries is the strip road. Here, the prepared surface is confined to two parallel narrow strips suitably placed for the wheels with again the need to drive off when passing. At the opposite end of the scale is the motorway style road which is a purpose-built limited access dual-carriageway with minimum bends, which avoids all towns and villages, has no frontagers and is without any conflicting traffic movements at intersections. This means that at junctions, flyovers or underpasses become necessary and in mountainous areas, such roads often involve tremendous feats of civil engineering. Construction is not, therefore, cheap. Suitably placed service centres are often provided giving facilities for rest and refreshment, but otherwise stopping is prohibited on these roads other than in emergencies on the parallel hardshoulder. Some of these highways in North America were built with the median strip wide enough to accommodate a special bus lane and in some cases these lanes have subsequently been converted to rail rapid transit. Chicago is an example where one of the newer lines is constructed in this way. All of these special motorways have at least two traffic lanes in each direction, most have three and, particularly in North America, some have much more than that.

Needless to say, these super highways have not sprung from bridleways. They are new roads and, along with all other new roads, they are built by cutting across open country with demolition if necessary. Some sections may even be elevated. But regardless of their importance in transport terms, there is an environmental connotation as the substantial tracts of land necessary for their construction does inevitably suffer a major transformation. Adverse public reaction is compounded by the fact that the demand for more road space is insatiable and the road becomes saturated with traffic as soon as it is constructed. People ask (that is, when they are not driving their cars), when is it going to stop? The way in the form of a public road is available for all to use. Land availability and the geography of the area will in the end dictate when no more roads can be provided and in the more densely populated areas that time is coming soon.

On roads generally, headroom is something that influences both construction and use and this varies by country. In Great Britain, for example, the standard headroom is 16ft 6in (5m) and this permits the operation of double-deck buses. In some other parts of the world but including Europe and North America the standard height clearance is generally less which can preclude the operation of tall vehicles.

As road vehicles are driven on sight, unlike the other modes, no sophisticated movement control measures are necessary. The occasional traffic lights, pedestrian crossings, warning signs and road markings are all that are generally required, although in heavily congested urban areas more elaborate forms of traffic control are being instituted. Systems of urban traffic control embody such measures as computer-controlled traffic lights over the entire inner zone network, the operation of which is optimized in accordance with the density of traffic flows on different roads and all linked to a central control room with visual displays on closed circuit television screens. Such a system adjusts itself to the traffic movements and any emergencies at the time and special priorities for buses can be built in. With the passage of time it is likely that vehicles will be guided by remote control along stretches of motorway and collisions avoided without assistance from the driver.

In built-up areas where space is at a premium, fast roads and urban motorways have been built on elevated structures. An overhead road provides a quick outlet for through traffic free of surface congestion and without headroom restriction and enables access into downtown areas. Such overhead ways could be several miles long or they could be short flyovers along roads to relieve conflicting traffic movements and hence congestion and delay at busy intersections. For public transport, provision can be made for stopping points with suitable access from the ground. On the face of things, therefore, elevated ways have much to commend them and they can be found worldwide. Elevated urban motorways have eliminated what would otherwise have necessitated extensive surface demolition as well as avoiding the barriers that the motorways would have created between the communities on either side.

Having stated the benefits of the elevated way, there is, of course, one fundamental disadvantage. It is visually intrusive. To those in the immediate vicinity it can have a claustrophobic effect: it deprives them of daylight; it is unpleasant to live with; it invades the privacy of the upper rooms of those buildings which flank it and it does nothing to abate the noise and emission factors of passing vehicles. In short, it ruins the environmental scene and similar remarks apply to the elevated railway as is mentioned below. But it is its attractions and benefits that tempt planning authorities to introduce elevated schemes, particularly where such outlets are desperately needed (if only to make way for the car!).

There now remains for consideration any special adaptations of the highway for the particular requirements of public transport. There are, of course, the special bus lanes and even bus-only roads but they are really only normal roads that have been subjected to traffic management schemes. The special applications referred to in this context are discussed below.

Guided bus lane

The principle of a guided bus lane is to set special guide rails or kerbs along a reserved bus lane for use by buses which are equipped with guide arms and rollers. This permits the bus to proceed with precision without steering assistance from the driver which in turn enables the guided lane to be narrower than that of a conventional bus lane. For example, in the case of a bus width of 8ft 2½in (2.5m) the guides need be set at a track width of only 8ft 6in (2.6m). Buses are, of course, free to leave the specialized track at any point where there is a suitable gap in the kerb. Examples of the system in use may be found at Essen (Germany) and Adelaide (Australia).

A guided track of this kind could also be used by trolleybuses (providing they are also suitably equipped) and tram tracks could be laid along it if required.

Trolleybus

Trolleybuses use the normal road surface but being electrically propelled require dual overhead wires for the vehicle's two poles for power collection purposes (there are no rails to complete the circuit as in the case of the tram) together with the provision of sub-stations for the electrical feed. Compared with the free-standing diesel bus, therefore, considerable capital equipment is involved and as was seen in Chapter 2, provision is made for trolleybuses only where traffic is sufficiently heavy to justify the expenditure. However, the dual-powered vehicle, also described in Chapter 2, does overcome this feature.

Tram

The conventional street tramway requires only tracks to be laid in (probably) the centre of an existing road (with rail that must be flush with the road surface) together with provision for the electric power supply similar to that of a trolleybus (but with only a single wire). It is interesting to note that as trams were one of the earliest forms of mechanical traction, parts of the original urban systems sometimes traversed short tracts of open country linking pockets of housing. As development continued, paved surfaces were placed around the tram tracks which then formed a part of the local road network. Now, the trams have gone but the roads remain in what have become busy dormitory areas and today probably few people realize how the road first came to be there.

At one time there were examples of the electric power being fed through a sub-surface conduit with a slot in the centre of the track which overcame the visual intrusion aspect; but the need to place the live rail in a hollow below the surface of the street increased the cost of construction.

It was said in Chapter 2 that to speed the passage of trams, it is advantageous to segregate the track away from the highway and on to a special reservation running alongside the road. This is now done wherever it is physically possible with the tram tracks impinging on to the roads only in places where there is no space to do otherwise (and then accompanied by special traffic management measures). It is worthy of note that if a special tramway reservation is paved (ie, not just rails laid on sleepers railway style) then it could also be used by buses.

LIGHT RAIL TRANSIT (LRT)

Although light rail transit (as described in Chapter 2) is here regarded as a separate mode, the infrastructure and vehicle requirements are a blend of tramways and railways suitably adapted for the purpose. No special attention need, therefore, be given to LRT in this context. There are instances where LRT systems have been planned with full heavy rapid transit (see below) in mind for introduction as the traffic develops, at which time, the signalling system, etc can be modified accordingly. This is known as 'pre-Metro'.

MONORAIL

A monorail consists of a continuous beam carrying a single running rail. The trains can straddle the beam and rail as in the case of the Alweg type or, as the line is usually elevated, be suspended from it which is the principle of the Safege system. If the track is elevated it must be supported on columns or suitable frames which, if the cars are suspended, must be sufficiently high to permit satisfactory clearance when crossing or running above and parallel with roads and railways. Switching at line bifurcations entails somewhat cumbrous equipment but if there is just one line with a terminal loop at each end the difficulty does not arise other than for a line into a car depot. Also, suspended trains are liable to swing in high winds and some control may therefore be necessary at station platforms.

The classic example of a monorail is at Wuppertal (Germany) where the urban area runs alongside a river. For most of the journey the route of some 8¼ miles (13¼km) spans the narrow gorge of the River Wupper and only for a short distance at the western extremity is it suspended over a busy street in a shopping and residential area.

CABLE TRAM

For gradients that would be impracticable to negotiate even by a rack railway, as is described below, cables may be used. The cables are set either between or beneath the tracks and are able to haul railed vehicles up very steep inclines with stopping and starting effected by gripping and releasing the moving cable in conjunction with the vehicle's brake. This system was originally one of the first attempts at propulsion without horses for routes along roads (which did not necessarily constitute abnormally steep gradients). Steam was the basic source of power to move the cable with the engine housed in a separate building. The concept was pioneered in San Francisco, USA, in 1873 and it spread to other areas including two short lines in London, England. But electricity came and the cable disappeared for normal traction purposes. However, it survived in San Francisco (where there are very steep hills) and a part of the system there remains. It is now the classic example of the street cable tramway which is operated along with the rest of the local transport services by the San Francisco Municipal Railway. Otherwise, cable cars are used not along roads but up hillsides, for example where a severe gradient precludes any other type of traction. Many examples of that type exist today.

The funicular railway is also, in effect, a cable railway. In this instance two vehicles of similar weight are linked by cable, one being at the base and one at the summit of the line. The cars balance each other as they ascend and descend with a passing loop midway if only a single track is provided.

CABLE CAR

A different form of cableway is that used in hilly regions either to ascend mountains or to cross valleys. In this case, cars are suspended from moving, aerial cables. In mountainous areas this mode has a transport value and there are extensive systems in the Alps in Europe, for example, which are heavily used in season by skiers and also by those who wish just to go and see the view, quite apart from fulfilling a need to get to otherwise inaccessible places.

RAILWAY

Having dealt with the infrastructure of the railed or fixed track forms of transport that are closely associated with the road or used only in a local or urban setting, we now consider the main line railway systems. The railway is a segregated way which gives it a potential for high speeds. But it is steel wheel on steel way and, as was seen in Chapter 2 in relation to trams, railed vehicles have a lower resistance to rolling, which was why a horse was able to pull a heavier load on a tram than it could on a bus. On the railway, however, this lower resistance to rolling (ie, less adhesion to the track) becomes also a handicap. It means that inclines must be less severe. The combination of these two factors (speed and adhesion), therefore, means that the track must be reasonably straight and level. The sharper the curves the lower will be the speed and gradients of 1 in 12 or less which are found on the roads are unsuitable for express trains. It follows that a railway must be as level and as straight as possible and cannot therefore adhere to the contours of the land as well as can a road.

The track gauge and the loading gauge are other inputs to the infrastructure specification; the track gauge being the width between the two running rails and the loading gauge the overall space required by the train according to the dimensions of the rolling stock in terms of width and height. In practice, the narrower the track gauge, the poorer will usually be the quality of the ride and it may reduce the maximum permissible speed, whilst the larger loading gauge can give greater capacity and afford better passenger amenities. But the wider the gauge the greater may be the minimum permitted curvature and the larger the loading gauge the more costly will be the capital works, particularly the tunnels.

With its rigorous requirements it is not difficult to see why land acquisition is necessary and why the cost of building a railway is high, particularly in hilly terrain. Indeed, it has led to magnificent feats of civil and structural engineering in the form of bridges, viaducts, tunnels and stations.

The specific width between the running rails is not crucial except for one thing. If different companies or different nations adopt different gauges then an operational problem arises when, on expansion, the systems meet. Through running, although desirable, is impossible unless the cumbersome and time-consuming procedure of changing bogies *en route* or using some form of wheel sliding along axle arrangement is adopted, and then this is only a palliative to what is an unfortunate situation.

Railways began in isolated pockets and not all had the foresight to adopt a universal track gauge. In Great Britain where the railway was born, the track gauge was set at 4ft 8½ in (1,435 mm) by George Stephenson when he laid his first railway in 1814 and it is alleged that this was for no better reason than that it happened to be the gauge of an old colliery track. However, it proved to be a satisfactory optimum width, became widely adopted and is now regarded worldwide as the standard gauge. Nevertheless, there are variants. For example, in Russia and the other countries which together constitute the Commonwealth of Independent States, in the Indian sub-continent, in a part of Australia and in Spain, Portugal and Ireland they are wider (but not all of the same wider width), whilst in most of eastern, western and southern Africa, Myanmar (Burma), Thailand, Malaysia and most of the western Pacific area, together with considerable mileage in India, Pakistan, New Zealand and again in Australia they are narrower. Australia is particularly unfortunate in the choices of its predecessors as it now has a variety of gauges, including the standard gauge, and is in the process of converting some lines to ease the complications. India and Pakistan have grown up with substantial systems of both broader and narrower (metre) gauges and there are stretches of track equipped with three rails to accommodate trains of both dimensions. Again, there have been some conversions. The Franco-Spanish border is an example where the break of gauge is inconvenient and it would have been unfortunate to say the least if England and France had different track gauges now that the two systems are linked with through trains operating via the Channel Tunnel.

It has been said that the railway can accommodate high capacity traffic at fast speeds even though there may be many conflicting movements because it has the sole use of its own specialized way and can enforce a strict code of operational discipline. That enforcement is effected through a signalling system which is all part of the infrastructure. In its simplest form, railway signalling is based on the block principle with the line being divided into sections (blocks), the lengths of which vary according to such factors as train lengths, frequency and speed. Each section is considered blocked until the signalman (or blockman) has obtained permission from a colleague further down the line for a train to proceed. The signals themselves were originally of the semaphore type with arms that were raised or lowered with illuminated colour at night to give the appropriate indication to the driver. As a train running at speed cannot stop at short notice, duplicating signals (known perhaps as distant or repeater signals) warn the driver of the state of the main signal ahead. If the repeater signal is clear then the subsequent 'stop' signal will also be clear and speed can therefore be maintained.

Such are the rudiments of early railway signalling and the system can still be found in areas where there has been no modernization, particularly in some of the less developed countries. But more than that is required on lines with very high speeds or high frequency services. Semaphores have been replaced by coloured lights and numerous lineside signal boxes have been replaced by computerized control centres able to embrace 100 route miles or more of track. The equipment is contained in a central control room and operated from consoles that contain illuminated diagrams showing the state of the signals and the positions of trains with push buttons and switches setting their paths. The visual indication to the driver can as an alternative be placed in the cab of the train rather than at the lineside. Then for high density lines there is automatic signalling where the trains themselves, through track occupancy, set the signal protecting the section to their rear at danger and which clears as they proceed. This is likely to be coupled with a built-in safeguard which prevents trains passing signals at danger. The ultimate is complete automation where the presence of a driver is not required but these sophisticated forms of control are applied more to the heavy rapid transit urban systems.

Enough has been said to see that modern railway signalling is expensive and highly sophisticated and is a complex science. It is a subject in itself and not one to be pursued here. But it is all part of the infrastructure and, however advanced may the technology become, there is one guiding principle that is inviolable: however close the trains may get to each other, they must always be kept apart.

One final aspect relative to railway infrastructure is of note. A railway is a statutory undertaking and in the majority of countries its permanent way and equipment and its methods of operation must comply with regulations and/or specifications as laid down by government. However, there are lines on which certain standards and legal requirements have been relaxed in return for specified operating restrictions. These lines do now tend to be disappearing as part of the conventional railway. However, over the years they have been designated 'light railways' which is getting closer to the more recent 'light rail transit' as was mentioned above.

HEAVY RAPID TRANSIT

In civil engineering terms, one of the most complex forms of transport infrastructure is that which involves underground railways. Constructed as they are in heavily built-up areas (they would not likely be underground otherwise) they involve deep-level burrowing through the subsoil with massive earth clearances at stations.

The earliest method of constructing an underground railway, however, was not by burrowing but by the 'cut and cover' method. Here, the road was dug up and then restored over the resultant trench thereby creating a shallow tunnel of sufficient dimensions to accommodate a standard sized railway. Just as Great Britain was the 'home' of the railway, it also saw the construction of the first underground railway (in London), the first few lines of which were built in this way; the first stretch was opened in 1863. But there is a limit to where this type of construction can be undertaken. Digging up a road is inconvenient enough. Demolishing property to any major extent is unacceptable and thoughts turned to burrowing instead of digging.

Again in London, the world's first deep-level underground railway was opened in 1890. Because of ventilation problems, it would not have been possible to run steam trains through tunnels of this type (steam trains ran through the shallow tunnels) but by the time that this burrowed track had become available, electric traction for trains had become a reality. The work of burrowing a deep tunnel (which is known colloquially as a 'tube') and attendant stations is a highly specialized science and it is not proposed to elaborate here on the engineering complexities of that operation. Suffice to say that it is accomplished with minimal impact on the surface and, with modern sophisticated methods of construction, tunnels do not now need to lie directly beneath roads as even the tubes did at one time. Tube lines are able to reach considerable depths if necessary to avoid obstructions such as other lines, water pipes, sewers, etc. As a matter of interest, the deepest station on the London Underground system below sea level is 70 feet (21.3m) whilst the maximum depth below the surface of the ground above (being under Hampstead Hill) is 221 feet (67.4m).

Underground rapid transit systems have now spread across the world and if a system has more than one line the chances are that they will cross each other. At these crossing points there will be interchange stations which in themselves are engineering feats of exceptional magnitude. To quote London again as an example, King's Cross Station has three separate deep-level tube lines under a shallower cut-and-cover line which calls for four sets of two platforms (one in each direction) all below the surface, in vertical succession, with the whole underground station being underneath a conventional main line railway terminal. Some measure of the magnitude of the construction can be assessed when it is remembered that eight underground platforms must be connected by passage or by escalator between each other and with the main line trains and the surface exits. Remember from the capacity estimates that each platform could accommodate trains carrying anything up to about 1000 passengers every minute during heavy bursts of traffic in the peaks, many of whom, at what is an important traffic objective and interchange point, will be either alighting or boarding. The walkways and escalators must therefore be of adequate proportions and numbers to accommodate this movement.

An alternative to the underground railway is the elevated railway. This is a cheaper way of doing things assuming that there is a way through to build the elevated structure but it does nothing for the view (other than for the passengers). This latter remark applies of course to any elevated way, be it urban motorway as already referred to, monorail or conventional main line railway.

Today, many of the large centres of population worldwide have a heavy rapid transit network. Any of the above methods of construction may have been used, even on different sections of just one line. It is likely that on the extremities of the systems where more space is available, yet another option will have been adopted, being that of a normal surface railway.

Another variant of the specialized types of way of fixed-tracked systems is that required by the pneumatic-tyred train. This type of train requires a track consisting of two parallel strips of concrete or other material that gives a good running surface for the pneumatic tyres of the train. Beside and above the running tracks must be two vertical strips which engage with smaller horizontal wheels attached to the train for guidance purposes on the same principle as the guided bus lanes (see Plate 36). The guide rails in this case, however, can be used also as the electrical conductor rails. Additionally, conventional steel rails are usually set alongside the concrete running tracks. These are used in conjunction with conventional steel wheels which duplicate the rubber-tyred wheels of the train. They can be lowered for use in emergencies should there be a tyre blow-out and also to guide trains through switches (points) and thereby avoid the otherwise cumbrous arrangement necessary for guiding the rubber tyres.

PIPELINE

Most major pipelines are buried in a trench with about 3ft (1m) of earth cover. Before being lowered into the trench, the pipe is likely to be covered with a thick layer of bituminous enamel and wrapped with a protective covering. In certain desert countries oil pipelines have been laid above ground. This is cheaper and reduces the problem of corrosion but owing to the large variations in temperature, allowance must be made for expansion and contraction. The physical obstruction caused by an overground pipeline, however, makes this method unattractive in cultivated or inhabited areas.

Pumping stations are necessary to 'propel' the commodity through the pipe and are, therefore, an integral part of the pipeline installation. There is a variety of prime movers such as steam turbines, electric motors, diesel engines, etc, and long lines have intermediate boosting stations. However, the technicalities of the overall operation controlling movement along the length of the pipe lie beyond this study. Suffice to say here that much of the flow can now be controlled automatically with built-in safety devices or by remote control with unattended boosting stations.

WATER TRANSPORT

Introduction

Seas and rivers are of course provided naturally as are some of the harbours. The man-made infrastructure consists mainly of the ports and navigating aids. Those rivers that are used for transport purposes have to be made suitable for navigation and it is only the canals that are artificially constructed. Again, the technicalities of the terminals required by ships is a study in itself and one which only those with a specific interest in that particular mode will wish to undertake. Nevertheless, there is also an environmental connotation and for that reason alone the subject cannot be ignored here.

Sea

The two fundamentals for a sea terminal are a harbour and a port. It has just been said that some of the world's harbours are the result of a natural formation but generally they have to be artificially improved with the aid of protective devices such as breakwaters. An adequate depth of water must also be maintained, both in the approaches and in the harbour itself, which could necessitate dredging. If this proves to be a continuous requirement it could become a heavy ongoing item of expenditure.

It was said in Chapter 1 that the locations of seaports go back into history when water transport was the first means of communication. For natural reasons, river estuaries were favourite sites. But as land transport developed, the ports became the transshipment stage of a through route and so the port requirements grew into something more than just what was needed for the ships themselves. There had to be land access as well as sea access. In addition to the rivers which were already there, railways and then roads became crucial parts of the port area and with them the need for facilities to load and unload, first manually, then mechanically and now in parts specialized and automated. With that came the need for storage areas, warehouses and provision for ancillary services such as customs facilities, ship repairs, bunkering, victualling, security, etc.

It is the port through which commodities are exported and imported; out from the inland manufacturing centres and in to the inland markets. The port, therefore, exerts a strong influence on life in the surrounding area on the landward side (or the hinterland). The size of the hinterland will vary according to geographical conditions and the relationships with other ports. It might be just a small surrounding area. On the other hand, the hinterland of the port of Colombo, for example, is most of the entire country of Sri Lanka (Ceylon).

Then there are the specialized terminals such as those of the roll-on roll-off ferries. Purpose-built pontoons with a vertical play to allow for varying tides are necessary to enable a smooth passage for vehicles passing between ship and shore. If it is a railway ferry then the differences in height are critical and in tidal waters the boat would need to be contained within a lock to stabilize the water level. Important in this context is a need for a large circulating area for cars, coaches and heavy commercial vehicles where they can assemble and wait in readiness to embark. Some of the boats are now quite large; for example, P&O's 37,583 gross tons *Pride of Bilbao* which plies between Portsmouth (England) and Bilbao (Spain) can accommodate up to 600 cars and 62 large haulage vehicles. When this is compounded by the fact that traffic is likely to be queuing for several different boats at any one time, it does mean that the area of parking space must be substantial.

Another development is the container port. Much of the liner trade network worldwide is now containerized and the demand for specialized ports has grown. Containers were discussed in Chapter 2 where it was seen that they represent a form of through transport. The ports do not, therefore, need to be as strategically sited as far as the hinterland is concerned as, regardless of the mode, the containers make their way through and beyond to inland ports and internal distribution centres. The infrastructure requirements, when put into deceptively simple terms but which nevertheless require careful planning, come down to physically suitable access from both land and sea; a large flat area of open space available for stacking purposes, suitably equipped with state-of-the-art gantry cranes with automated handling equipment.

Inland waterway

Not all areas lend themselves to canal construction. As for obvious reasons a stretch of water can only be level, a hilly terrain is not canal friendly. Furthermore, the soil must not be too porous and there must be an adequate supply of water. These circumstances will not only determine the practicability of canal construction, they will also condition the width and depth. Holland, for example, being low-lying and flat is very suitable for canals but England, which has a more hilly terrain and does not possess such an abundance of natural water, is not as appropriate and canals there are restricted. In other places, canal construction is impracticable.

Clearly, canal territory does not have to be billiard table level (if it did there would be no canals); there are means of overcoming differences in height. The usual method of doing this is by the construction of locks between different levels of water. A lock is, in effect, a short section of waterway contained within sluiced gates and its operation involves filling the chamber with water from the upper level which then slowly drains down to the lower level and the vessels within the chamber rise or fall according to the direction of travel. Unfortunately, the use of locks makes what is already a slow form of transport very much slower. Passing through a lock is time-consuming and the filling process means that the canal must be replenished with water. The hillier the terrain, the greater is the need for locks even to the extent of installing a flight of locks with one immediately following another. This explains why some countries can have wide, deep waterways able to accommodate large ocean-going ships and which are important commercial transport arteries, whilst others have little to offer and some none at all.

An alternative to the lock is the lift. Here the vessel floats in a cradle of water and the entire load is lifted or lowered as the case may be. This is a more expensive way of doing things but substantially faster.

AIR

The air, like the sea, is a natural way. But with dense commercial airline networks spanning the world with large numbers of aircraft in close proximity to each other, converging on airports at fast speeds, there must be a sophisticated system of navigating aids and elaborate terminal facilities. To put it in perspective, consider 2 planes 10 miles (16km) apart cruising at 600mph (960kph) but approaching each other on a collision course. If they remained on that course it would take only 30 seconds for them to collide. There can be no 'driving on sight' in those circumstances even in daylight and clear weather. An air traffic controller must guide them and keep them apart, particularly on take-off and landing. Remember that unlike the other modes, an aircraft cannot stop in flight. It must continue to fly and to remain airborne means maintaining a minimum speed (and along specified paths). However, to give some idea of the magnitude of the task, at a large busy airport, aircraft with anything up to about 400 passengers apiece (if they are B747s) can be queuing in the sky waiting for permission to land, maybe at a rate of perhaps 1 every 90 seconds. With air traffic control being so highly complex and specialized, its examination becomes a subject beyond this study of the principles.

The provision of suitable terminals is also crucial to air travel but again something that cannot be considered in depth here. Nevertheless, a broad reference will be made to airports and they do have environmental repercussions.

Airports can be quite small. That at Daintree in the north of Queensland in Australia, for example, has just a simple grass landing strip with a windsock to show which way the wind is blowing and it has a park-type bench seat for four people beside some bushes in an otherwise open field which is designated 'departure lounge'. At the other end of the scale are the large international airports with multi-runways and multi-terminals covering thousands of acres (or square metres) of land and with a throughput of some 50 million or more passengers per annum.

The basic requirements of a large international airport are:

- one or more runways of sufficient length and strength to accommodate the aircraft that use them;
- facilities for air traffic control including the control tower;
- adequate road and ideally also rail access;
- facilities for feeder transport, eg, bus stands, railway station, taxi ranks and car parks;
- adequate arrival and departure bays for aircraft and access thereto including passenger access, assisted maybe with travolators, internal automated railways, etc;
- facilities for aircraft maintenance, refuelling and general servicing;
- terminal buildings adequate for both incoming and outgoing traffic with all passenger necessities and amenities such as check-in and baggage collection, waiting accommodation, food outlets, shops, customs, immigration, passport control, police, security, etc.

As well as all this, the airport needs to be as close as possible to the town which it purports to serve but, as was said in Chapter 2, that is often not very near. A very large flat open space is generally hard to come by in a built-up urban conglomeration where in any case low-flying aircraft are not very popular. Although passengers do not view favourably a long trek to the airport (and remember from Chapter 2 that there are a lot of tourists), the noise of aircraft overhead is not acceptable to residents who live under the flight path. Emission from aircraft is also serious, particularly as most of it is at a height where maximum damage is incurred. It is a feature of airports, therefore, that their commercial and environmental attributes are very different.

The Structure and Financing of the Industry

INTRODUCTION

Transport infrastructure or, in other words, the capital works and equipment necessary to provide the services, was dealt with in Chapter 3. The first part of this chapter is about the various bodies which are involved in the provision of transport and which collectively constitute the structure of the industry. It is about forms of ownership and how the services may be provided. Attention was drawn in Chapter 1 to the fact that by and large transport was pioneered by private enterprise. Over a period of time many of the larger undertakings were absorbed by local or national governments and now there is a swing back to private ownership. The one branch of transport that has remained relatively free throughout from state participation is shipping but even here there are numerous instances to the contrary, all in accordance with local politics. Public ownership does tend to produce larger units which in turn has a bearing on internal organizations which are discussed in Chapter 5. It is the main forms of ownership within each of the private and public sectors which are to be considered here.

In this particular study it is important to remember that the type of ownership is only a means to an end. It should not influence the science of transport, the principles of which are the subject of this book; transport management (which can always be divorced from ownership) is a profession in its own right regardless of the form of ownership. Having said that, the priorities of the private and public sectors could be different. The entrepreneurs must live and they must service their capital. They could not survive otherwise. It is inevitable, therefore, that the profit motive looms larger in their thinking and the provision of any poorly patronized but socially desirable services will be limited. But to maximize their profits they are more likely to run their businesses with minimum cost and maximum efficiency. Although some are, it is unfortunate that managements in the public sector are not always so motivated. However, with politicians in direct or indirect control and with access to public funds, service provision might be considered before profit and that side of it could be beneficial. Note, however, that there is a tendency, particularly in the developing countries, to overemphasize the distinction between the public and private sectors, almost to the extent of regarding them as separate and distinct modes, which is quite wrong. Ownership, management and mode must not be muddled.

Within industry generally, ownership of business units appears in different guises and most of the various forms are represented amongst undertakings which provide transport services. However, the differences in types of service and territories served are so great that it is impracticable to generalize on the merits or otherwise of any one form of ownership. For example, a bus service into new development or a single (maybe domestic) air service could be provided satisfactorily by small-scale enterprise, but as the area expands and population increases so the demand grows to the point where the greater resources of a larger organization (publicly or privately owned) might be required. A railway undertaking, if it is of any size at all, will clearly fit into one of the larger classifications. But in transport generally, organizations of all types and sizes have their place. Nevertheless, transport attracts the attention of politicians very easily and this also has a bearing on the structure of the industry. The ultimate pattern could well be for political rather than for financial or professional reasons.

The formation of partnerships and companies is governed by legislation which will vary country by country. This is not a legal textbook and even if it was it would not be practicable to elaborate on company law as it applies in every part of the world. What follows, therefore, is only descriptive and outlines some of the different types of ownership that are to be found in transport in approximate ascending order of size. Each can provide a valuable service to the community in its own particular sphere, within its own natural limitations and within the legal framework of the country concerned.

Having considered the various types of ownership both in the public and the private sectors, this chapter will then go on to examine the parts played by both in the provision of capital to finance the infrastructure that the transport undertakings require and to see how any new works can be justified.

OWNERSHIP

Sole proprietor

Sole proprietors are in complete charge of their vehicles and equipment and accept entire responsibility for running their businesses although they could employ labour to assist them in their work. Their assets may have come from personal savings or more likely from borrowed capital or through leasing agreements, the cost of which must be redeemed out of income. In addition to their managerial responsibilities, it is not uncommon for the proprietors themselves to take an active part in the more menial tasks and their staff could consist of their families whose remuneration might be something less than normal rates. Their office, if they have one, could be a room in their private house in which only the most rudimentary system of accounts might be kept. This does not, however, mean that the sole proprietor is inefficient or produces an unsatisfactory product. In reality it might be quite to the contrary. Depending on their attitude, people who live on the success of their endeavours might not wish to maximize profits at the expense of quality and goodwill. In any case, minimum standards may be required by law. Nevertheless, observance of the law tends to vary with individuals and with the degree of enforcement. In some parts of the world bribery successfully overcomes this particular inconvenience.

Because their equipment must be provided from their own resources, sole proprietors will seek that branch of transport which requires minimum capital investment. Road transport is therefore the obvious choice although a small ferryboat could be another example. With lower overheads and maybe cheaper labour, operating costs will be less and for this reason, sole proprietors can often succeed where the population and consequently the traffic is sparse and insufficient to support the heavier expenses of a larger concern. But the financial resources of an individual are limited. Expansion of this type of business is not, therefore, easy and owner operators who seek a wider field might combine their resources with those of another and form a partnership.

Partnership

The term 'partnership' defines the relationship that subsists between persons carrying on a business in common with a view to profit. Two or more persons can by this means pool their resources and together they will be in a stronger position to expand. The arrangement could also bring in additional expertise. Otherwise, the characteristics of a partnership are similar to those of a sole owner. They can still be hampered by a lack of working capital and the extent of operations, therefore, remains limited. Like the sole owner, partners retain their own personal identities and they have an unlimited liability for the debts of their enterprise. As the undertaking expands, so the money involved gets bigger and the risks become greater. This feature is important as it could be costly to the owners if the venture fails.

Although speculation involves risk, people will be loath to invest if the consequences of failure are catastrophic. To lose money is one thing; to incur substantial debts and possible bankruptcy is another. If enterprise is to be encouraged, therefore, some better arrangement is necessary.

Company

As has been noted, legal issues and hence the finer aspects of company law (and even the terms that are used) vary country by country. For example, as the participants of legally registered companies enjoy certain privileges (and this applies once their undertaking is suitably qualified and has been accepted for incorporation on the register), the term 'incorporated' now has a familiar ring in the USA, whereas elsewhere, undertakings are more likely to contain the word 'company' in their title. But for the purpose here, it is just the principles that will be discussed.

Undertakings within the company form of ownership range from the small to the very large. The dictionary definition of a company is, among other things, a 'body of persons combined for a common (commercial) object'. In one sense, therefore, any organization except that of a sole proprietor could be regarded as a company. In this context, however, the company has a special meaning. Whilst the literary definition is a general one, the legal interpretation is more precise, as an undertaking must comply with certain statutory requirements, whatever they may be, before it can be legally incorporated on the register as a company. These requirements are designed to ensure that as far as is possible the company is reputable and are likely to cover, for example, compliance with obligations related to capital structure, rights and duties of directors and members and the preparation of accounts.

In this context, a company is a distinct entity. The members subscribe towards a predetermined level of capital necessary for floating the business in return for which they hold shares and receive dividends or interest in accordance with the type of shares held and the prosperity of the enterprise. It is much easier than in the case of a partnership for individuals to withdraw. If they so wish, they can just sell their shares. If the company requires more capital, then extra shares may be issued. More importantly, a limited company gives protection to its shareholders to the extent that, should the business fail, their individual liability is limited to the extent of their holdings (hence the term 'limited'). Possible hardship that could otherwise be suffered by a sole proprietor or members of a partnership is thereby avoided and most corporate businesses are structured in this way. Shares in the company are not necessarily available for the general public and if not it is known as a private company. However, in the case of large-scale enterprise it is likely that the company will wish to invite public subscription. Its shares would then be listed on the stock market, at which time it becomes known (in British parlance) as a public limited company (plc).

In the transport industry the company style form of ownership embraces an enormous range of different undertakings ranging from quite small to the very largest multi-national giants such as, for example, the international oil companies, that operate within the private sector. This type of structure does not, however, necessarily indicate private ownership. It is now but a short step to public ownership. Participation by government in the purchase of a proportion of the shares produces a mixed economy (part private and part public ownership) with control dependent on the extent of the holdings. When government holds all of the shares it then, of course, becomes public ownership but still within the company structure with its own independent management. If, therefore, transport undertakings are to be in public ownership, the company type of organization can still be retained or introduced as an alternative to the other possibilities that follow after the co-operative, which is something quite different.

Co-operative

A co-operative is an organization that has been created as a result of individuals banding together and contributing their own private resources into what is a common venture. The people concerned (the members) may then take an active part in running the business and each will receive his share of the profit. All members will have an equal interest in the prosperity of their undertaking and hence a responsibility to ensure that it functions satisfactorily. If the resultant unit is sizeable, then an internal structure on the lines described in Chapter 5 becomes necessary; this means the creation of a management team with departmental administrators and local supervisors down to drivers and mechanics. However, regardless of the job that they are actually performing, assuming that they all have equal shares then they are all equal participants in their co-operative. The various tasks are therefore allotted from the members by the members according to their aptitudes and qualifications. With a body of equals all working for a common cause, questions of status and grading levels do not really arise and all are likely to receive similar remuneration. Any individual malpractices detrimental to the running or goodwill of the service would be dealt with by a committee drawn from the members as would major policy decisions. The workforce could be supplemented by paid employees who would be controlled in the same way as staff in any other type of organization.

The thought of equal status regardless of the work being done may at first seem strange but with all participants having an equal interest in a common venture and all endeavouring to make it work, the system does have logic. It is something that is not widely practised but a classic example is in Israel where there is a bus co-operative serving the Greater Tel Aviv area and another covering the remainder of the country.

The system should not be confused with worker participation in management where a member of the staff (probably a trade union representative) is elected also as a part-time director and attends board meetings, or with management/employee buyouts as sometimes happens in the privatization process. Certainly in this case employees are allocated shares with voting rights on major decisions but management is otherwise in the hands of the executive directors.

Readers may be more familiar with co-operatives in the retail trade. This, however, is likely to be different as the staff in this case may be paid employees, whilst the membership consists of the customers who receive a share of the profits without participation in management.

Public corporation

Between the publicly owned limited company with its loosely defined responsibilities and direct administration and control by a government department where the head of the organization is a minister of state or a chairman of a committee of a local municipality, lies the middle course of the public corporation. This type of undertaking is created by special statute which details its rights and responsibilities. Machinery is thereby available to allow privileges and to enforce more stringent conditions than is possible in the case of limited companies, such as the provision of a certain level of service in prescribed areas or financial limitations and obligations. It is likely that a public corporation will not be able to raise private capital and any national restrictions on public sector borrowing powers (and in times of financial stringency there may well be) can severely restrict the undertakings activities and its ability to develop and meet adequately public demand.

The theory of the public corporation is based on the separation of management from ownership and the undertaking itself is, like the company, a legal entity. It has perpetual succession and a common seal, it can sue and be sued and has a degree of autonomy not found in government departments. It has its own workforce, the staff are not government employees and hence are not classified as civil servants. According to its constitution, a government minister would be answerable to parliament on policy matters and overall performance but internal administration would not be subject to his direct control.

In the public sector this type of organization has merit and is appropriate even for the largest undertakings. The arrangement became popular around the 1960s. However, there are alleged difficulties in raising capital and with the current obsession for privatization this is sometimes held as a reason (or perhaps an excuse) to attain what is really a political ambition. As it is governments who fix the rules, it is not abundantly clear why governments cannot also change the rules to allow public corporations to borrow money if it was practicable so to do. The objection is that those in authority, although not elected by the people, would be able to speculate with public funds. On the other hand, they might, perhaps, be able to raise private capital. Whilst not wishing to labour the point, it is felt that there are circumstances where the problem might be more apparent than real.

As enterprise is always credited to the private sector with the now popular assertion that privatization must be the trend for the future, it is opportune at this point to quote a study in Karachi (Pakistan) which the writer recently undertook. At that time, Karachi was served by some 6000 private sector buses (mostly midibuses) under the control of almost as many different owners. There was also a public sector bus corporation which was able to carry only a small percentage of the total traffic but with standard-sized single deck buses. In so doing and unlike the private sector, it incurred heavy deficits attributable to excessive administration and general lack of efficiency. It has been said that all transport should be run by professionals in a professional manner and that the form of ownership should not affect the results, but here in Karachi it did, with the costs incurred by the private sector being much lower. However, a further look revealed the following:

- The public sector observed the labour laws, the length of drivers' hours and wage agreements but the private sector did not.
- The public sector provided its own off-street overnight parking but the private sector did not.
- The public sector had purpose-built facilities for vehicle maintenance but the private sector did not.
- The public sector had to comply with government directives made for political reasons and which militated against economical operation but the private sector did not.

There were also other differences but these alone are sufficient to indicate that if comparisons are made, the circumstances must be similar.

This is often not done. Like must always be compared with like and it not always is.

Municipality

From the middle course of the public corporation comes direct owner-ship and control by a government department, in this instance local gov-ernment. A municipal undertaking functions as a department of a local authority. Elected members of the council form a committee that deter-mines policy and the management and staff are local government employ-ees. Note, however, that government administration varies between countries. Below central (national, federal, etc) government there will be various lower echelons such as states, provinces, counties or whatever, down to local districts. In this context, the term 'municipality' is a some-what broad reference to government other than central national govern-ment, but as the definition of a municipality is a city or a town the inter-pretation is generally regarded as appropriate more for the smaller con-fined territories. There are many examples worldwide of local authorities owning and directly operating (through their paid employees) bus and tram networks, ferries, sometimes local metro systems, seaports and air-ports as well as, of course, parts or all of the public highway system and other items of transport infrastructure.

Note that the company form of ownership can also apply in the munic-ipal context. It was said above that governments can own the shares of companies and there are examples where municipalities also own the shares of local transport undertakings. Nevertheless, under these circum-stances such organizations operate independently of direct local govern-ment (ie, political) control.

State department

The last form of ownership to be mentioned is that of complete owner-ship and direct management by national or federal government with the undertaking functioning as a department of state.

In this case, overall responsibility, even down to day-to-day running, is vested in a government minister who employs his own staff who are clas-sified as civil servants. This type of control, however, tends towards a pol-icy geared more to the discharge of ministerial responsibility to parliament than to a commercial outlook. So often it is neither sufficiently dynamic nor properly focused on running a business that needs to attract cus-tomers and remain competitive in a commercially hostile environment. It is for this reason that state participation in any such enterprising activities was in some countries changed over the years to the principle of the pub-lic corporation (and then privatized) which has been described.

Overview

Looking at the situation worldwide, all of the above mentioned forms of ownership are represented in the transport industry. A policy of public or private ownership is a political issue. Some governments (but the number is diminishing) do still see public ownership of all national and socially necessary utilities, including transport, to be desirable, regardless of circumstances, and they see the public sector undertakings as instruments for implementing their own strategies. The prevailing climate, however, is now veering towards privatization. With some governments it has even become an obsession. They seem prepared to go hell-bent along this path regardless of any practical complications almost as if that particular political goal is the end in itself and not just a means to improve (in this context) the transport system. Of course, finance (or the lack of it) has a lot to do with it.

FINANCE

The structure of the transport industry is very much bound up with finance and the provision of finance for the infrastructure (ie, capital investment) will now be considered. Much of public transport, at least on the passenger side, does also receive revenue support but as that may bring with it a measure of control, that side of things is dealt with in Chapter 7.

The modes which involve heavy capital expenditure on infrastructure are air (for airports and navigation aids) and rail (for track and stations). Shipping is disregarded in this context as passenger services are very limited and the provision, modernizing and equipping of seaports for the throughput of merchandise is not as much in the public eye and is likely to involve private capital anyway. Both airlines and railways play an essential role in the provision of transport services, albeit for different types of journeys, but it is the passenger traffic on the railway which, because of what can be very heavy capital expenditure, is seldom able to live just on its revenue. At best that revenue would likely cover only its direct costs. But it is a national asset and, certain new lines could be of strategic importance or might be necessary to regenerate a depressed area. Remember from Chapter 2 that in the urban transport scene, it is the railway that can move the masses but the capital cost of deep-level tube lines are particularly high. Some special attention must, therefore, be given to the railway in this respect.

Whilst road, sea and air operators all share the use of their way and/or their terminals, railway undertakings have traditionally been the sole owners and users of the entire system, being the track, signalling, stations and rolling stock. There are now instances, and Great Britain is an example, where upon privatization, these constituent parts of a hitherto national system were divorced and have hence become separate independent organizational and accountable units. In Great Britain, for example, the rail operating companies (there are now some 25 of them, sometimes with more than one using the same sections of track) lease their rolling stock from independent leasing companies and their drivers drive the trains observing signals and on track owned and maintained by yet another company (in this example – Railtrack). It is appreciated that airlines do just that at airports whereby planes take off and land in accordance with the directives of an independent flight control authority using airports which are owned and operated maybe by yet another independent authority. In transport terms, therefore, the concept is not new. Nevertheless, in the case of the railway, the various assets which collectively form the 'railway business' are seen by the writer as being more appropriate for a single organization. The services run more naturally as an integrated unit and, in the minds of the users, the stations and the trains have a single identity. A divided ownership can really only create timetable compilation difficulties with different operators requiring use of the same piece of track at similar times, railway staff working side by side but for different employers with few outside people knowing who their employers actually are, station staff having responsibility for only some of the trains without any knowledge of the others, complex administrative procedures whereby train operators pay for the track as they use it, etc.

To digress for a moment; an undertaking whose expenses include a high portion of overheads can normally employ the marginal cost theory when, for example, planning special facilities or additional trains for traffic promotion purposes. Although slightly out of context, let it be noted here that the marginal cost theory is something which is based on the premiss that with the overhead costs already met, there is usually room for a 'little more' with the extra expense being confined to the direct costs of operation. In other words, only the avoidable costs would be taken into account (the unavoidable costs would have been paid anyway) when costing the provision of, say, special cheap tickets or an excursion train. This gives scope for enterprise which is not there if a proportion of the full track costs have to be taken into account and perhaps calculated on a mileage basis. However, so much for this diversion. Attention must now return to the more direct financing of the infrastructure.

It is submitted that although the principle of divorcing the operator from the infrastructure is the same for both air and rail, in practice the circumstances are different. With so many different airlines, most of which are foreign to any one country, to do otherwise would be impracticable. That is not the case with a national railway system and there are disadvantages in this splitting of the assets. The same reasoning applies to the train services themselves. With many journeys requiring interchange no purpose can be seen in allocating them to different operating companies, particularly when interworking, connections and through tickets are called for. On the other hand, the compensating advantages are far from clear and the writer sees no logic in this arrangement. Nevertheless, the issue is a relevant part of both the structure and the financing of railways and as such it has to be mentioned.

Having said all this, however, other than noting the separation of the railway assets in Great Britain, we have so far got nowhere with the consideration of public or private financing of the infrastructure. As far as the railway is concerned, it is usually virtually certain that any investment will not be profitable in purely accounting terms and hence produce a financial return. Along with that, however, important projects require a large amount of capital often beyond what national exchequers will gladly support. In this case thoughts have turned to the possible injection of private capital to enable new works to proceed and there are instances where this has been done. An example again lies in Great Britain with the extension of the Jubilee Line in London. The Jubilee Line is one of the lines of London Underground (a subsidiary of the publicly owned London Regional Transport) which is being extended across east London to serve the newly developed Docklands area (a one-time shipping port but now no longer used as such) where a new rapid transit facility is of strategic importance. But the new line runs alongside and under the River Thames through ground consisting of clays, silts and sands making it rather like digging through a wet shingle beach. Deep-level tube construction (as referred to in Chapter 3) in this area is not only a feat of such complexity that it could not even have been accomplished until more recent times, it is also exceptionally costly. The problem of funding came and the result was a mixture of public and private capital. This, of course, begs the question, why should private speculators invest their money in such a venture, knowing that such capital will never be serviced from passenger fares? The key to the answer lies in other benefits.

Cost-benefit studies

The retention and development of a public transport network might well bestow an overall benefit greater than what is measured by an operating deficit. It is in a cost-benefit study where these other external factors (which can never be found in the accounts of the transport undertakings themselves) are considered and evaluated. Reference was made in Chapter 1 (and will again be in subsequent chapters) to the private car, its impact on the environment and the increasing imbalance between available infrastructure and capacity demand. The resultant cost through congestion, which a good public transport system would reduce, is an example of what are known in this exercise as 'externalities'. A cost-benefit study involves an evaluation of such externalities to bring them into a constant relationship with the commercial costs and benefits, being the normal revenue and expenditure cash flows. In other words, a cost-benefit analysis is an attempt to set the total costs to the community of a particular project against the total benefits, including the social benefits, to the community. In this way, the net advantages can be judged. Those with overall responsibility such as governments and planning authorities are then in a better position to make rational decisions to the maximum advantage of the people generally. Decisions of this kind could not and would not be taken by those with only sectional interests.

In the world of transport, the railway, with its high capital cost but often high return, if not to itself, to the community, is an appropriate subject for cost-benefit analysis. A railway administration that is charged with responsibility for running its system as a commercial venture will look only at its own finances when considering which lines it should develop or what facilities it should provide. From the railway viewpoint, domestic costs and revenue would decide whether, for example, a line should be either electrified or closed. For the electrification option, the considerations would be confined to just those extra costs and revenue that would be attributable to the change of traction and in the case of closure, to the marginal savings that would be achieved set against the loss of revenue. In other words, only those costs and benefits that are directly involved would be taken into account. The same principle applies to a cost-benefit study except that instead of just sectional railway interests, it is the community interests that are evaluated. Considerations that are brought into a cost-benefit analysis are again simply those elements that are directly related to the particular project and benefits that would not otherwise be realized but in this case related to the community rather than just to the railway. The construction of an underground railway, for example, would relieve highway congestion and save expenditure in other ways such as the provision of new roads and the implementation of road widening schemes. There would be repercussions on existing transport services, it could influence traffic patterns and flows by encouraging more people to travel along certain corridors and, indeed, into the centres of particular towns. Land values in the vicinity of stations would rise.

But stop there. It is at this point of the discourse where the topic can be related back to the previous section and which explains why private capital might be injected into such things as underground railways. It is because people will come, trade will come and developers can then see the value of their property appreciating to an extent that the cost of helping the railway to come, that all this is justified.

The same principle applies in reverse should a railway line be considered for closure. Over the years there have been many such closures worldwide but in the light of events as will be considered in subsequent chapters, maybe some of those closures might not have been in the best interests of the longer term. The railway might have lost a service and with it an operating deficit but the permanent loss to the area, if it could be quantified, could be very real. But there is more than one way of providing communication and cost-benefit studies must evaluate all reasonable alternatives whereby the declared objectives might be achieved, covering both capital investment and regulation such as parking restraints and heavier fuel tax. To encroach on to Chapter 9, the implementation of either or both of those last mentioned options must now become a very real possibility which might make the retention of a railway even more important. This of course only goes to emphasize the care and foresight that must be fed into studies of this kind. Such a study involves a general economic and physical appraisal of the relative advantages and disadvantages of possible options prior to a detailed analysis of one particular solution; again only those factors are considered that would not take place but for the implementation of the project in question.

A cost-benefit exercise is dependent upon predictions and requires a price to be set on the options that emerge. Evaluation of the externalities is a particular problem and one that distinguishes it from other financial approaches. For example, it is known that there is road congestion and that it involves cost to the users when, for instance, drivers and their loads are kept idle and hence unproductive in traffic blocks. It is a cost-benefit analysis that requires a monetary value to be attached to all costs and all benefits, including those that have no market price and cannot easily be quantified. The resultant figures will not, therefore, be precise but the principle is there. As far as transport is concerned, it takes it out of its isolation to become instead a part of the social environment.

Management

FUNCTION OF MANAGEMENT

It is the duty of management to work within the policy directives of its governing body (which, as has been seen, may be a government department, a board elected by shareholders, etc, according to the form of ownership) and at the same time to comply with the requirements of the statutes. To discharge its responsibilities and achieve its end product (which in transport is a safe arrival) management will need to delegate and to do this the administrative machine must be organized into a suitable structure according to the type and size of the undertaking concerned. Top management has responsibilities to its governors, to the legislature, to its customers and to its staff. At local level and if the business is sizeable, a tier of middle management will be sandwiched in between with responsibilities to its senior management, to its passengers (or consignors of merchandise) and to its workforce.

The purpose of business generally is to remunerate capital and create wealth. Management is required to conduct its affairs accordingly in keeping with policies as laid down; its principal functions being to plan, to organize, to motivate and to control. This is what top management is all about and the ability to do these things is an essential attribute of the chairman, the president or whoever represents the undertaking at the top and who carries overall responsibility. The necessary skills (in this case transport) are then represented on the management team, being the board of directors. Even if the owner and manager are one and the same person, as is the case with the sole proprietor, they are still separate roles, one taking the risk and the other remunerated to make it work. It is the owners or business venturers (or entrepreneurs as the economist would say) who, through their vision and drive, see justification and scope for a business and it is they who launch it. But it is the function of the manager that is now for consideration.

Although to a large extent industry depends on these entrepreneurs as it is they who seek out new business, they may have done little more than identify that demand. Unless they are specialists in their own right they will need professional managers to develop their enterprise who in turn will also need to exercise their skill to exploit the potentials. The point being made is that those who risk capital do not necessarily have a detailed involvement in the business and may only participate in the formulation of broad policy; entrepreneurs need not directly oversee the achievement of their objectives and it is management who undertakes the ultimate task. It is likely, therefore, that management will be required to do more than work within a set of rules contained within rigid and detailed terms of reference. Management at the top requires also drive and initiative, which in transport means an awareness of what is needed, keeping abreast with new developments and giving the customers the service that they want whilst at the same time keeping within the constraints of its governing body. In short, they must also develop their policies.

BASIC CONCEPTS

It is a task of management to determine where it is going and how it is going to get there. Policies must, therefore, be formulated. In transport, some undertakings have a statutory basis, that is, they have been created by government which, in so doing, has laid down their rights and duties. Governments usually have their own political transport policies and this reflects in the types of organizations that they have either created themselves or have allowed to come into being. It is then for management to produce its own policies as it deems appropriate to meet its obligations. Its responsibilities are as follows:

- to owners (or governing body) who require either maximum return on their capital or a satisfactory provision of facilities at minimum cost according to the terms of the constitution and within any legal constraints;
- to the customers who require adequate and satisfactory facilities at the cheapest fares or charges;
- to the staff who require congenial working conditions with maximum remuneration.

Added to this, mention will be made in Chapter 7 of a process of tendering either through a controlling authority responsible for the provision of a co-ordinated network or where socially desirable but unremunerative services need to be provided. In this case there will be a fourth party to whom management has a responsibility: a government-appointed body who invites and then deliberates on the observance of the terms of the tender.

The whole process resembles three (or with tendering, four) thorns pricking into its side and management must plot its tortuous path through this difficult scrubland. Note that it is top management that has contact with the governing body but it is usually the lower echelons that have contact with the customers and the day-to-day workforce. They are in on the ground floor and it is for them to reconcile the irreconcilable with their own top management bosses. Management at all levels is in the centre of opposing forces.

Management must identify and analyse its problems, establish its objectives and then look into the future, make forecasts and produce a corporate plan. It is this process which demonstrates its skill and is the key to success or failure. Corporate planning means planning the forthcoming business of the company, covering every activity within the organization, as well as taking into account all relevant external factors. This therefore involves all internal functions whatever they might be. The strengths and weaknesses of the undertaking can then be assessed, objectives quantified and suitable strategies determined to facilitate achievement of the declared goals. A master plan designed along these lines should cover at least the ensuing two or three years but probably also, in less detailed form, the following, say, ten years after taking into account all likely trends and fortunes that are expected to take place in that time. This of course means that management must be dynamic in its thinking and long-term plans must remain sufficiently fluid to allow a continual process of revision and updating. With the passage of time and as events unfold, programmes materialize and sights are rolled forward on to new horizons.

OBJECTIVES

It has been said that the end product of transport is a safe and satisfactory arrival. To achieve that arrival money must be spent on capital equipment to varying extents according to the mode but, in any event, this requires investment and the owners will require a return on that investment. Transport undertakings could, therefore, impose a business objective of, say, a certain percentage of profit growth per annum and policies would be developed accordingly. They could also prepare declared objectives to achieve certain standards of service, new facilities, etc, perhaps as a part of normal business trading or maybe to meet certain statutory obligations.

As an example, the five-year 1997/98 to 2001/02 Corporate Plan of the State Transit Authority of New South Wales,[1] Australia (which is a public sector statutory body charged with the responsibility of operating bus and ferry services safely, efficiently and in accordance with sound commercial principles and which operates buses and ferries throughout Sydney and Newcastle) may be cited. The five key strategies which form the basis of this Plan relate to:

- operations/services
- finance and capital expenditure
- staff relations
- ecologically sustainable development
- regional development

which are supported in the business plans of its three operating divisions: Sydney Buses, Sydney Ferries, and Newcastle Buses and Ferries. The plan indicates the Authority's strategy which is, *inter alia*:

- introducing new services to meet the changing needs of the community;
- pressing for infrastructure improvements to speed the services and making them more comfortable and accessible;
- achieving independence from Government funding for operations with all subsequent returns being invested into further improving services;
- making the services as attractive as possible to entice people away from private cars;
- promoting what are its environmentally responsible services and purchasing environmentally friendly equipment;
- providing bus and ferry services in a manner to promote regional development (in Newcastle).

The Authority is conscious of its requirement to provide the best possible services. The point is made that the latter half of the twentieth century has seen a dramatic growth in cities and that both Sydney and Newcastle are now more dispersed with lower population densities. Interestingly, the Plan goes on to say that this has encouraged more cars on to the road and hence an environmental cost through traffic congestion and air pollution. The Authority then claims that if things are to be environmentally sustainable, an efficient public transport system is an essential component to enticing people away from cars which in turn encourages consolidation and limits urban sprawl. It then considers that the greatest contribution that it can make to the environment is by increasing bus patronage and claims that every bus (they are single-deck with standees) has a potential to save up to 80 separate private car journeys, thereby improving air quality (and road space). The Plan indicates that buses with the highest standard low emission diesel engines are already in service and many of the latest deliveries use the environmentally friendly compressed natural gas (CNG – see Chapter 8) which is an abundant natural resource in Australia.

Note that this represents a situation which is mirrored in many of the larger urban areas worldwide and is all very pertinent to the comments and proposals that are contained in Chapters 8 and 9.

The Authority's programme for the five years from 1997/98 include among other things:

- reducing costs to commercial levels;
- continue the fleet replacement programme;
- install a new automatic fare collection system on the ferries;
- develop new cross-regional bus services;
- investigate new ferry services to support harbourside developments.

Whilst referring to the New South Wales Transit Authority this is a convenient juncture to note also that its new ferryboats are of the twin-hulled catamaran type which have a low wash and are faster than the older style conventional displacement hull boats. The new high-speed craft are better able to service a commuter market that has become more time sensitive as well as a growing leisure market. A conversion of the ferryboats to CNG is also currently under examination.

This is just one illustration. Such decisions reflect the character and style of an undertaking and its function and place within, in this instance, the transport industry. The formulation of sound objectives and policies in order to formulate a corporate plan is an essential ingredient of good management just as is the need for the long-term planning of networks and facilities referred to in Chapter 1. This is a subject which fits conveniently into either this chapter or Chapter 1 as there is a close relationship between management objectives and planning proposals. Other examples of corporate plans are contained in Chapter 1.

MANAGEMENT RELATIVE TO SIZE

The size of an undertaking has an influence on its internal organizational structure, as will be considered below, but it also affects management styles. There is a limit to what an individual can effectively oversee. It follows that in large undertakings, senior management must delegate to a greater extent than in small ones, which underlines the importance of the ability to lead and motivate. It also requires, as was mentioned at the outset, an additional tier of middle management. This can mean, however, that top management then becomes remote from the customers and the workforce whilst middle management is distant from the owners (or governing bodies). In a large undertaking, therefore, there is an added duty to minimize the effects of this situation with adequate communication at all levels. But as an undertaking gets bigger, so the administrative machine grows.

To illustrate the effects of administration within a large organization, suppose more information is required on such things as the effects of fares revisions or changes in services or working practices and a new section is created for that purpose. The cost does not lie only with the extra people that have to be engaged for that purpose. It will become somebody else's responsibility to recruit and train them, to find accommodation for them, to compile and retain their staff records, to take care of their canteen and other welfare requirements and to administer any pension rights. In the interests of 'efficiency' it will be yet another task to see that the jobs were justified in the first place. Amid the jungle of people and paper, situations can arise where, for example, a clerk prepares regular information which, possibly because of a change in circumstances, no longer serves any useful purpose; but because it is nobody's job to stop it, it is passed on to its accustomed destination where another clerk is employed to file it. Procedures such as this are particularly rife in the public sectors of the developing countries where the administrative arrangements are often so overwhelming that they have a negative effect and become a delay factor. Sometimes it is not discouraged if only to keep people in work. It reduces unemployment, labour is cheap and maybe at the end of the day someone else is paying. But it certainly does nothing for efficiency. Nevertheless, in terms of administration, some of the large nationalized units have tended to became top-heavy, even in the West, with plenty of room for weeding. Further reference is made to this aspect under the sub-heading 'The human factor' below. Even so, the economy of size does not apply to transport to the extent that it does in some other industries.

Having said all this, no lessons have been learnt on what might be the optimum size in management terms of a transport undertaking. Really there is not one. It is a question of what is most suitable for the prevailing circumstances, bearing in mind the various factors that have been outlined. All that has been attempted is to try to bring out the deficiencies of being either too big or too small. Transport developed from small units. Over the years and for a variety of reasons, many became welded into large, probably publicly owned, monoliths. Some have been broken up in the process of privatization perhaps only to grow together again in the private sector. Finality will never be reached but large-scale enterprise does have a history in this respect. Without wishing to suggest fragmentation, over-administration is something about which all concerned should be on their guard.

ADMINISTRATION

Introduction

Having introduced the basics of management, next for consideration is how undertakings arrange their internal organizations in order that they may run their businesses in a workmanlike manner. As this is a book about transport, it is transport undertakings that will form the background to the examples. That apart, however, transport is a very suitable industry to use as a base for the discussion of organizational principles: it embraces authorities with compact networks, others whose activities are far flung, perhaps even spanning the world, and those that have scattered areas of traffic concentrations possibly long distances apart, not to mention the great variety of sizes in terms of areas served or traffic moved. An examination of the subject even in just the transport context will therefore bring out a range of administrative arrangements.

Organizational principles

Like forms of ownership, a system of organization is a means to an end. A successful organization must utilize the resources of the undertaking to produce maximum efficiency and economy. Different levels of responsibility must be defined and understood. There must be sufficient flexibility to permit prompt decisions and adequate machinery to ensure that they are properly executed. Some predetermined form of organization is therefore essential. There must always be control but as there is a limit to the number of people whose work one person can effectively manage, delegation becomes necessary and duties must be allocated. A chain of command must therefore be established.

The functions of a manufacturing concern fall into three groups. The raw material must be bought, it must be processed and the finished product sold. Each one of these functions is a necessary attribute to successful business and the failure of any one would lead to collapse. In transport, this theory remains; the fact that the finished product is intangible makes no difference. Raw materials (vehicles, vessels or aircraft together with other capital equipment) must somehow be obtained and, like machinery required for manufacturing purposes, must be kept in good working order. Once the carrying units become available they must be judiciously programmed to provide (or 'manufacture') the most appropriate public transport service. At the same time, that service must be sold and the price must be determined in relation to costs, thereby ensuring that the undertaking remains viable.

In transport terms these groups may be broadly identified as:

- buy and maintain (engineering)
- make (operating)
- sell (commercial)
- viability (finance)

which between them conduct the day-to-day running of the system. As in any other business, there must also be provision for general administration and legal matters which could all come under the jurisdiction of, for simplicity, say, a company secretary. These various activities must be organized or, in other words, put into proper relationship. Translated into a 'family tree' type diagram, the organization of a small or medium-sized undertaking would in its simplest form appear as is shown in Figure 5.1. If the five departmental heads were considered too many in relation to the size of the unit and the work involved then the operating and commercial managers might be combined into, say, a traffic manager. If on the other hand it was too large for the five to embrace then the engineer, for example, might be split with, say, a mechanical engineer and an electrical engineer, and an administrative officer would supplement the secretary. All would report direct to a managing director, general manager or whatever. These titles are commonplace in road and rail transport. In sea and air transport, different terminology might be used. For example, operating and commercial might be referred to as movement and sales but the functions and principles involved would be the same.

It is not sufficient in the process of organizing merely to identify the inherent parts; those parts must work together in a manner most suitable

Figure 5.1 Organizational chart – the functional system.

for the particular circumstances. There are many variations in the way of doing this but the structure of any undertaking stems from one or other of two systems, namely the departmental or divisional chains of command.

The skeleton chart at Figure 5.1 might be developed into a departmental (or functional) system. However, there can be dangers in departmentalism when applied to large-scale enterprise. Large departments are prone to act in the interests of only their own particular activity rather than of the undertaking as a whole of which they are but a part. Also, time can be wasted in committee and when departmental representatives (in a large organization they can all be quite high powered) battle in support of sectional interests the result can be a bad compromise instead of an informed decision. A strong functional system therefore might bring with it the danger of producing a series of different spheres of influence working without proper reference to each other. Human nature is such that under this arrangement senior staff are liable to see themselves as being the supremos of autonomous departmental units rather than as part of a team. The larger the undertaking, the stronger will be the departmental chiefs and the more weighty will the supporting headquarters staff become. Formal contact between departments becomes tedious with day-to-day matters travelling up through a top-heavy administration, across a 'bridge' (a representative committee or someone very senior with authority over more than one department) and then down again the other side.

Even with the dangers so described, there are examples of this type of organization in practice. Where the size of the undertaking permits management to keep close to the ground or even in large units where activities are sufficiently stereotyped and/or disciplined to make local discretion less necessary, the system is satisfactory and can be commended. But where departments become too large and powerful in their own right and in what should be a dynamic industry where quick on-the-spot decisions are necessary to meet the challenge of competition in an ever-changing pattern of demand, it is the person on the spot, not at a remote headquarters, that requires executive power. It is the divisional system that better meets this situation.

Think again of the embryonic structure. Departmental heads were responsible for their respective functions right down the line. Consider now the position as it would be if these departmental chiefs were replaced by local managers responsible directly to whoever is at the top for all day-to-day matters within localized specified areas. Note the reference to day-to-day matters. In a situation such as this, it might be desirable that certain aspects of management should still be dealt with centrally, such as, for example, accounts, legal issues, overall planning and maybe publicity. With provision made for this, the skeleton chart would now appear as shown in Figure 5.2. This chart, which represents the divisional theory in its sim-

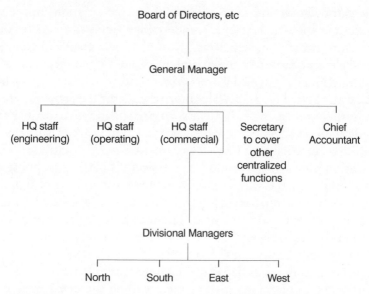

Figure 5.2 Organizational chart – the divisional system.

plest form, assumes that the operating area is split into four divisions, in this example, north, south, east and west. Such divisions could be geographical regions or districts or perhaps lines if there were two or more main trunk routes within the system. The important point is that the divisional, regional, line or district managers, whatever they might be called, are genuine managers responsible for all matters local to their respective spheres of influence and which would otherwise be in the charge of separate departments. They would be subject only to policy directives from headquarters. One person would, therefore, be responsible for obtaining the traffic and moving it. A reduced headquarters staff would be responsible for policy and general co-ordination together with those functions reserved for the centre. The feature of the divisional system is that it brings management right in on the ground floor and gives to the local people much of the responsibility hitherto reserved for headquarters. Certainly the local managers would need to have functional specialists on their staff but being very much smaller units, the adverse features already described should not arise.

The divisional system does to a large extent overcome the problems inherent in departmentalism but at the same time it brings with it other disadvantages. The various regional units might like to see themselves as separate and distinct enterprises linked only by perhaps a common logo and vehicle livery (if that). Furthermore, whilst with the functional system there might be a lack of communication between the different departments, the problem could now become one of duplication of functional staff with conflicting edicts issuing from central administration and local managers. This could give rise to animosity between these two parts of management. Looking again at Figure 5.2, another less satisfactory feature is that whilst the functional staff in the divisions owe allegiance to their local managers (in this case designated 'divisional managers') the entire workforce must abide by the policies as determined (on behalf of the general manager) by the functional chiefs at headquarters. This has the flavour of working for two masters which is never ideal.

What has been said covers the broad principles of the two systems that form the nucleus of the many variations in the organizational structures within different undertakings. There are no rules as such and the ultimate pattern that any one company may adopt is no more than a development of the thinking described above, modified as necessary to meet the specific requirements of the business. Some have even adopted a blend of both but then there is the danger of achieving the worst of both worlds. Even this, however, is something that has not been unknown.

The trend towards privatization and smaller units should have had a slimming effect on some of the administrative machines. Although, beware: all of the units need top managers and the more units there are the more managers there will be. But as a further slimming exercise it has been found more economical to contract out certain functions which hitherto constituted departments manned by the company's staff. There is logic in this as in the opinion of the writer it is really only the operating specialists within the hierarchy as described who are the true transport professionals (or 'transportants' as people responsible for providing transport facilities are sometimes referred to). This is not to decry the other functions; they are crucial to the running of any transport system. Nothing would work without them. But the engineers, accountants, solicitors, caterers, publicity agents, etc are all qualified professionals in their own particular trades and they could equally well apply their skills elsewhere. It can only be the work of the operating or movement specialist that has no application in any other field although members of the other professions could of course be 'transportants' as well. But to return to the point: in practical terms, some work could be contracted out, in which case the actual company workforce would be much smaller.

It is suggested that the specialized tasks on the operating (or traffic) side include:

- route and service planning
- timetable planning
- vehicle and staff schedule compilation
- running or operational control

- long-term planning
- fare levels ⎫
- fare collection ⎬ (in conjunction with the finance function)
- special facilities
- publicity ⎫
- public relations ⎬ (in conjunction with the advertising function)
- staff control
- vehicle body design (in conjunction with the engineering function)

Even then, it is likely that schedule compilation and revenue collection will now be confined to preparing specifications for passing on to trained computer programmers and electronic machine experts.

If the above list is accepted as an approximate guide to the sort of things that are germane to the traffic department, with the other functions in this arbitrary classification being on the periphery, then it does give a guide to the possibilities and extent of what might be feasible to contract out in a slimming exercise. Note also that for air travel and sea cruising in particular, the great bulk of bookings, fare collection and ticket issue is already contracted out in the form of travel agents (with tickets often sold at a discount). Longer distance passengers by road and rail may also book and pay for their travel in this way but, as in this case, and particularly by rail, tickets are often bought at stations immediately prior to travel, in percentage terms the number of tickets that are sold through agents is less.

The human factor

At one time transport was a labour-intensive industry. In the developing countries where wages are usually still low and conditions of service poor it still is. But in the West where staff remuneration is higher, a large workforce has become costly. Furthermore, unsocial hours make recruitment more difficult. Developments over recent years have therefore been directed towards reducing the number of employees which has meant replacing men by machines and by automation. Examples are loading and unloading of goods (particularly at container ports) which has been automated, advance reservations have been computerized, the crews of large seagoing tankers have, with the aid of computers, been reduced substantially as also has the number of air crew on the flight decks of aeroplanes; bus conductors have disappeared, railway trains have less staff and where rapid transit systems are automated do not even need drivers, station platform staff have been replaced by closed-circuit television, bus and tram tickets can be bought in advance from ticket dispensers or in shops, office equipment such as electronic calculating, tabulating and other machines have replaced clerical staff, and so it goes on. With privatization, managements may have become even more cost-conscious with a deterioration in pay and conditions and yet further staff reductions. It is unfortunate but perhaps inevitable that one effect of all this is that the morale of some of the remaining staff is low. This can only put in jeopardy efficiency and loyalty to the service even though the eroded pay and conditions are still substantially more attractive than were those of their forebears.

It is this last observation that brings into focus the general improvement in wages and conditions that have taken place over the latter half of the twentieth century in accord with the higher living standards which the populace in general now seems to regard as its right. This is something which the transport industry (amongst all others) has had to bear but in this case it has been borne in the face of falling traffics. Prior to this enlightened age of industrial relations it is true that the working population in general and transport employees in particular suffered long hours for little pay under severe conditions and discipline. The interests of the shareholders (in the form of dividends) and the customers (in the case of transport in the form of cheap and efficient services) were upheld at the expense of the staff. In other words, labour was exploited and this engendered distrust between employers and employed. Certainly the record of management in this respect is not without blemish. Harsh decisions were made but at the same time the goodwill of any service is dependent on high standards and those high standards were generally achieved. Although few can now have any working recollection of life before and up to the 1930s and any feelings amongst the present labour force, true or distorted, must be the result of 'education' rather than experience, mistrust has prevailed and it is unlikely that privatization will do anything to diminish this lingering resentment. This is to some extent responsible for much of the industrial legislation that came in the 1970s. The relevant statutes in Great Britain, for example, cover such things as employment protection and redundancy payments, health and safety, equal pay and opportunities, sex discrimination, race relations, etc, to say nothing of the Social Chapter of the European Union. All of these things add to costs and tie the hands of managements in the hiring and firing process so often necessary for operational efficiency and economy.

This is not a book about industrial relations and there will be no elaboration on the finer issues of resource management. But the human factor has had an influence on the disappearance of some of the mammoth publicly owned undertakings which in turn has affected the structure of the different units. Further attention must, therefore, be given to this particular aspect and some background knowledge of the events of the 1970s is desirable.

The legislation of the time gradually transferred the balance of industrial power from the employers to the employed and although it is Great Britain that is here being taken as an example, similar situations prevailed elsewhere. This includes the developing countries where these circumstances are particularly applicable to the larger organizations within their public sectors.

The better wages and conditions which followed with worker participation in management did not produce better relationships with management or even worker satisfaction. They produced instead an insatiable demand for more and to the users that meant more expensive and less efficient services. The trade union movement came to be associated with friction and was generally regarded as the root cause of an unsatisfactory situation. Constant rounds of high wage demands, restrictive practices with resultant overmanning, go-slow tactics which some saw as 'working to rule' (it might have been better if there had been more working to rule), lightning and official strikes, picketing, closed shops (recruitment restricted to trade union card holders) and other related activities, all contributed to this reputation. On top of that, the legal hurdles which were imposed and which have to be surmounted to sustain dismissals for unsatisfactory conduct also undermine the administration of discipline and hence the quality of work. Certainly a fair day's work deserves a fair day's pay and 70 years ago it was the fair day's pay that was in question. But in the 1970s it was the fair day's work that had become the doubtful factor. What was still more unfortunate is that conscientious workers who continued to place emphasis on service and take pride in the results of their endeavours sometimes found themselves exposed in a hostile environment as they became unpopular with their 'brothers' within the trade union movement. This was particularly so in the larger undertakings.

One of the most blatant examples of trade union power and its challenge to management authority was in the British ports. In the early 1970s the erstwhile National Dock Labour Scheme (which was a scheme introduced twenty-five years earlier to meet a hardship arising from casual labour) was extended to reinforce the permanent status of what were known as registered dockers. The bodies set up to regulate employment in each of the port areas involved and to carry responsibility for maintenance and allocation of an adequate workforce, together with recruitment, training, discharge and discipline, were known as National Dock Labour Boards and these boards had a 50 per cent representation of both employers and trade unions. This, then, was not just worker representation on the board or joint consultation; it was joint management. It so happened that this arrangement developed in the era of new technology when ports were being modernized (reference has already been made to the replacement of men by machines) and the permanent status of the docker began to run parallel with what was becoming a diminishing labour requirement. As a result, under the joint management regime, men were being paid to do what was no longer there for them to do. They were being paid to do nothing, which in practice meant going home (or 'bobbing', to use the colloquial term). Furthermore, the docks became a breeding ground for restrictive practices; if specialist personnel were brought in to work new and sophisticated equipment, they had to be 'ghosted' by an equivalent number of registered dockers, again being paid to do nothing. All this, of course, was a serious deterrent to healthy development and good only for driving away investment and patronage. It was not until the principle was finally abandoned in 1989 that the ports concerned (it was not all of the British ports that were involved), along with the economy, began to prosper.

Although the problem in the British ports in the 1980s was an extreme case, it was clear that the pendulum had swung too far. It is accepted that over the years and up until around the 1930s labour had been exploited with miserable pay and conditions. The trade unions had struggled hard, first for legality and then for recognition, as was exemplified by the prosecution of the Tolpuddle Martyrs over 160 years ago. It was understandable that the workers would not now easily surrender their newfound but hardwon power. But it was unfortunate that with power did not come responsibility. They had become greedy and even the term 'worker' came to be regarded as a mockery. Something had to be done.

So this was the labour situation which was nurtured within the large public sector organizations. But could all of the blame for this unsavoury situation really be put on to the shoulders of the big bad trade unions? The answer must be no. Consider now the situation, still in Great Britain but this time not in the ports but on the buses in the major conurbations at around the same period. It has already been said that before organized labour became a force to be reckoned with, managements were harsh and paid scant regard for the welfare of their employees. The workers could not therefore be blamed for thinking (correctly) that if the unions went away those same conditions would return. And just as the trade unions through their restrictive practices, etc brought about the need for larger workforces, so did managements jump on the bandwagon and enlarge their already substantial internal administrations. The principal functions as already described were expanded as opportunities were found to introduce new departments. There was 'empire building' and staff establishments grew. New breeds of experts appeared on the scene such as strategic planning officers, integrated services officers, economic efficiency officers, traffic development officers, organization and methods officers, executive planning officers, financial investigation officers, manpower services officers and productivity officers, to quote but a few. The rank and file found whole new layers of middle management in their midst which prompted a reversal of the old adage: 'Lesser fleas have little fleas put on their backs to bite 'em; little fleas have bigger fleas and so ad infinitum.'

But the lower levels came as well. All of these departmental heads required their own supporting establishments. A big house has many servants and there must be servants to serve those servants. Administrations mushroomed and with it the salaries of those at the top to accord with the numbers of people under their control. Trade unions may have been greedy but so also were managements. Something had to be done.

Then there were the governing bodies; in this case local politicians as it was local government that was primarily involved in this particular saga. Some made statesmanlike decisions and helped bus services to be maintained in places where they would otherwise have disappeared. But there were also some heavy spenders with taxpayers at their mercy. There were examples of policies determined not for professional reasons but to satisfy political dogma. There was an instance where for political reasons fares were held down at an inordinately low level and below what the traffic could bear. The result was a tremendous deficit for taxpayers to meet. This angered central government as it was considered, rightly or wrongly, to be for reasons of defiance, local aggrandizement and votes at the next election. Trade unions and managements may have been greedy but so also were the governing bodies. Something had to be done.

These events have been recorded only to illustrate to the reader the sort of things that might but not necessarily do happen in large-scale organizations. In the event the whole structure of the industry was subsequently changed. It could well have sparked off the privatization process which has now become worldwide and in Great Britain it encouraged de-regulation (on the buses) which is considered in Chapter 7. The changes were severe, perhaps made more so by a central government incensed by the activities just described – and they have been described here because they are germane to the human factor which is what this sub-section is all about. But the resultant reorganizations are essentially a part of the structure of the industry which was discussed at the beginning of this chapter. Something had to be done but only time will tell whether the resultant side effects are worse than the disease.

Total transport control

This final sub-heading on the subject of management within the transport industry also has a bearing on the structure as was discussed in Chapter 4. It is about separate, and on the face of it, independent functions working together under a common umbrella, probably in the form of a locally elected, political body. It is mentioned here as it is a system which has found favour in some places and there are examples of this arrangement both in practice and proposed.

The main application of this system lies in the provision of local transport where the (probably local) authority has responsibility for numerous transport and transport-related functions such as town and country planning, roads, traffic, car parks, bus operation, etc. The theory is that with common policies and goals, under a single manager, all of the relevant functions can then work as one and so produce schemes to their mutual advantage. As the supporting staff are drawn from each of the disciplines, the level of co-operation which is brought in at the top can percolate down to ground floor level and any privileges needed by, for example, public transport, such as for bus priorities, can be included in planning proposals at the embryonic stage.

On the face of it, this system of inter-departmental or multi-discipline co-operation does have logic and the protagonists claim that it can smooth out joint problems with buses and road space before they arise. But beware; it is a 'two-edged sword' and certainly is so as far as the transport side of things is concerned. Whilst the personnel actually involved might work well as a team, their deliberations would likely be (and in this instance certainly would be) subject to political control. Although (to quote a possibility from this example) the provision of bus priorities might be desirable (more likely necessary), the imposition of too many car restraints will never be popular politically and loss of popularity might mean loss of votes at the next election. This means that a planning team could be given a policy directive either before they even start or when initial proposals are submitted. At an early stage and at a low level, therefore, the transport element (or any of the other elements for that matter) could have lost its voice and from that point onwards is silenced and hence is unable to make its case. Professional viewpoints on the separate functional issues would then never reach higher authority (in this example, possibly central government) with the danger that the wrong policies are implemented at the whim of local politics.

The lesson here is that the professional views and recommendations of each separate function, in this instance the transport function, should be publicly heard, in its own right, through to the highest level of the ruling hierarchy.

Consumer Affairs

INTRODUCTION

This chapter is about the particular features of transport which are of special concern to the customers, together with the type of machinery that exists to protect their interests; the customers being either passengers or consignors of merchandise. It is cost, quality and the extent to which facilities are provided that are always uppermost in people's minds. There now follows a discussion on these issues with commentary as seen through the eyes of the users. The attraction and retention of a satisfied clientele depends on these things.

Looking back at Chapter 5, it is the commercial or traffic function which usually has contact with the users as it is that department which is likely to carry the main responsibility for planning the services, determining fares and charges and any special excursions, looking after the public relations side of things, dealing with the press and preparing publicity and any other advertising matter. It will assess what services or facilities are needed, ensure that the products are in accord with demand and then sell them. Within transport the function involves market research, planning, price fixing, publicity and the establishment of good customer relations (goodwill). All of this can be bundled together and identified by the contemporary term 'marketing' with business stimulated by special traffic promotion measures.

FREIGHT CHARGES

Preamble

If people are to use public carriers instead of making their own arrangements with their own private transport, charges must be attractive. This means that there is often a fine balance between the minimum revenue that a carrier must have and the maximum level of charges that the user will pay. There are various methods of rate fixing and to understand the principles involved, some elementary knowledge of both economics and finance is necessary.

To the economist, the theory of price fixing is based on the law of supply and demand. Demand, however, cannot be properly considered without reference to price because, in a free market, if the price rises the demand falls and if the price falls the demand increases. Then on the supply side, in a period of plenty, prices will be low and in times of scarcity prices will be high. In a truly competitive situation (which in transport seldom arises), it is the interaction of supply and demand freely operating in an open market that determines the price. In transport, one of the few examples where this theory can be seen in practice is the Baltic Exchange (which is described in Chapter 7) where the charter price of ships is directly influenced by the availability of suitable tonnage and the extent of demand at the time.

Cost of service

Disregarding the natural economic forces just described, the cost plus profit basis is the simplest and most easily understood method of calculating charges. It is a basis which is adopted by small businesses in road haulage and also in larger concerns for the charter of special vehicles, etc for use as a whole. To prepare a figure, proprietors estimate their direct costs such as wages, fuel, tyres and a proportion of maintenance based on time and distance, add a percentage to cover overheads and a further percentage to give a reasonable profit. The total is the basis of the charge. With all that done, the operator might still accept a lower rate such as on a return trip when the vehicle would have to run in any case or to encourage a regular customer and thereby maintain goodwill.

If the resultant quotation is competitive then that particular carrier might get the job. If it is not, there must be thoughts about either lowering costs by making the business more efficient or by narrowing profit margins. This illustrates the point that the greater the supply the greater the competition which in turn forces prices down. Whilst it was said in Chapter 1 that because transport is a highly perishable commodity this is not necessarily a good thing, it does apply more to passenger traffic where services are mostly pre-scheduled. In the case of goods where a requirement is more often met by a special arrangement made in accordance with a pre-booked order, the degree of wastage is not the same and this side of the industry can, therefore, be more competitive.

Value of service

All that has been said so far relates to a straightforward set of circumstances. But things are rarely that simple. Moving away from the small-scale road haulier, there is not far to go before it becomes impracticable to apply the cost plus profit theory to each separate transaction with any degree of accuracy.

There is, however, another feature that can never be disregarded when calculating charges. Thought must always be given to what the traffic can bear. The function of transport is to move goods (or people) from where they are to where they are required to be or, as the economist would say, to where their relative value is greater. Raw materials at their source have little or no intrinsic value (ie, no utility or use) although they have value in exchange. But they have greater value to a manufacturer when in his factory and the finished product has still more value to the wholesaler after it has been transported from the factory. Similarly, the wholesaler must dispose of his wares to a retailer and the retailer to the customer where, at this final stage, values increase to the extent that they are then available to satisfy a want. At each stage the goods become more valuable and it is the difference between these values at the start and finish of each journey that represents the worth, or added value, of transport. It is transport that has supplied the utility of place and this is the basis for setting the price for transportation of the goods. If the cost of transport exceeds the added value of the consignment as a result of its journey, then that cost is beyond what the traffic can bear and the service will not be used.

The theory of charges based on the value of the service to the customer or 'what the traffic can bear' is an alternative concept on which to draw up a rates book. The system does have realism. It began on the canals and was later developed by the railways at a time when they carried the bulk of the freight traffic. The system is suitable for these forms of transport because:

- Overheads form a high percentage of total expenses which makes it virtually impossible to calculate the cost of any one transaction.
- The heavier the traffic, the greater is the spread of overheads. Hence, the cost of movement varies in inverse proportion to the amount of traffic carried to a far greater extent than if direct costs predominate.
- Railways and canals are bulk carriers. The units are large and can usually accommodate 'a little more' without seriously increasing costs.

The significance here is that marginal traffic, even at a lower rate although not involving much cost, makes some contribution towards overheads.

Rates based solely on the value of service concept pay scant regard to distance. This is not entirely illogical as transport costs do not increase in direct proportion to the length of journey. But neither does the system have regard for loadability which was described in Chapter 2. Loadability does of course influence costs which, together with the distance aspect, does suggest that the system might be something less than perfect. In fact, if it was applied in its purest form it would be fraught with anomalies (as it has been) as different materials would bear different rates even for the same journey. Returned empties, for example (where they are still sent), bear little appreciation in value at their destination compared with their full loads on the forward journey but the cost of transit could be much the same. Without allowing for the cost element it would seem also to be a bit hit and miss financially; in fact almost a case of hoping for the best when drawing up the final accounts at the end of the year. And then what happens if there is a deficit? The system has been used by 'common carriers' (where there are any), being a carrier who is obliged by statute to carry anything anywhere and to accept all traffic as it comes. The railways in Great Britain, for example, were classified as common carriers at one time and did run into financial problems when competing with road haulage which was not so inhibited.

Clearly, a system of charges based on what the traffic can bear must also have some built-in refinements and reservations to take care of the shortcomings and dangers as described. But this can be done and the principle is an important element to be taken into account when rate fixing. However, with such an infinite variety of potential goods and journeys, a centralized rates structure with a multitude of material classifications quickly becomes unwieldy and impracticable. It is something better left to local managers on the spot to quote according to circumstances prevailing at the time and this has a strong bearing on the internal organizational policies as were discussed in Chapter 5.

PASSENGER FARES

The theory of cost and value of service as just described is still there in the background when setting passenger fares. Whilst on the cost side it is impracticable to calculate the cost of conveyance of any one person on any one journey, totals can be used on a route or system basis which can then be averaged out on a per-passenger mile basis. Total revenue (including any subsidies) must then exceed total costs if the undertaking is to remain operational. On the value concept, operators will not delve very deeply into the different values of people at their points of origin and destination but the passengers themselves will, if only subconsciously.

People travel for business or in accordance with their own personal arrangements. If, for example, a man could earn an extra £60 per week in a neighbouring town but the additional cost of transport was £80 per week, then, in the absence of any other benefits, he would not travel. (He could still take the job by moving house or buying a bicycle.) If the cost was an extra £75 per week, he might still not be interested but if it was £50 per week, he could be. His cut-off point beyond which he would refuse to travel would, therefore, lie somewhere between £50 and £75 per week, being his perceived additional worth created by his presence in the other town. An attractive shopping expedition at a fare of £15 return and his wife would not go but at a fare of £8 she would be tempted. She might still choose to go at £10 or perhaps even at £11 but for this non-essential trip (ie, the demand is 'elastic') she would not contemplate it even at £14. This means that if this particular potential passenger is to make this particular journey the fare beyond which she would not travel is something more than £11 but less than £14. That would be the cut-off point. Any more and then, in her estimation, the cost would be more than the benefit that she would derive. The fare would be beyond 'what the traffic can bear'. But it is not just one person involved. Everybody is a potential passenger and the higher the fare that is set between the realistic parameters, the greater is the number that will opt out. The resultant figure will depend at which end of the scale the majority begin to fall away and somewhere along the line there will be an optimum maximum charge.

It has just been said that the demand for this shopping trip is 'elastic'. It is non-essential. There are shops nearer and, in any case, there are other ways of passing the time including staying at home. But passengers are committed to their travel to and from work. It is more essential and is hence classified as 'inelastic'. These people think deeper and accept tougher conditions before they decide not to travel. Resistance to price rises will be less and operators can set higher fares with greater confidence that they will retain this type of traffic. There is not the same need to tempt commuters with special promotional measures as they do the optional rider. But remember always that for any type of traffic, if a fare is pitched beyond what the traffic can bear, it is tantamount to providing no service at all as the facility will not then be used.

Once the required total revenue is established, basic fare scales can be determined which will now be briefly considered. However, in all of the modes, special fares with a still stronger flavour of 'what the traffic can bear' are available. These will be discussed later. There are also different fares for different classes of travel if more than one class is provided. Again, this feature is discussed separately.

There will invariably be single (ie, one-way tickets) plus, perhaps return (ie, there and back tickets), unlimited journeys between two points for a given period (ie, season tickets) or anywhere anytime within a defined area (ie, travelcards, etc), to quote the more common examples. For inter-urban journeys and beyond, distance invariably has an input to the final charge but for local journeys in urban areas a flat fare regardless of distance is becoming more widely adopted. As a compromise between a scale based on distance and a flat fare comes the zonal system which is sometimes used in very large urban areas where a flat fare is impracticable.

Whilst fare levels are set with a revenue target in sight, policies are determined also in conjunction with fare collection. In the case of goods, the cost of carriage is invariably paid within the confines of an office (or home) and the act of actually handing over the money is not a physical problem (even though it may be a financial one). Similar circumstances apply to passengers when travelling by air and sea where the journey is booked in advance. For travel by train, payment might not be quite as easy as tickets are usually purchased at the station shortly before departure and with large numbers of people involved there may be delay and a scramble but not sufficient to prejudice the issue of a range of different tickets. For regular travellers, prepaid season tickets overcome the problem. On urban rapid transit systems, electronic turnstiles at stations in which passengers insert their tickets at each end of the journey have done much to reduce fraudulent travel.

It is on local buses, trolleybuses and trams within busy urban areas that the real problems of fare collection arise. A crush of people all making short journeys means dozens of separate cash transactions in a short space of time on the vehicle in difficult and cramped conditions. In the Western world where labour is expensive, this work is even undertaken by the driver which slows down loading. At the same time, the undertaking must receive its rightful dues. This means that all passengers must pay the correct fare and drivers must hand in all of the cash that they receive. Under such circumstances, fare collection methods must have a bearing on policy and the more fares that are paid for off the vehicle in advance of travel the better it is for operational efficiency on the road. Without delving deeply into the subject, examples of the more widely used systems are as under:

Fare collection by driver with a ticket machine	Time-consuming, particularly if there is a requirement to give change
As above but assisted by prepaid travelcards	A substantial help providing a sufficient number of travel cards

	are sold. Not easy for the driver to check the small print indicating availability
Machine inside vehicle for passengers to pay their own fare	Fast but relies on honesty of passenger and best for a flat fare
Farebox for passenger to drop money into a locked vault and visible for the driver to check	Fast but ideal only for a flat fare. Generally no change is given
Prepaid multi-journey ticket for driver to cancel as appropriate	Reasonably fast but ideal only for a flat fare

None of the above systems is perfect and all have both advantages and disadvantages. The ideal has not been attained. In the above options it is difficult or impossible for the driver to check that passengers do not take a longer ride than for what they have paid or even that the prepaid 'display' tickets are valid. Passenger irregularities can be discouraged but not eliminated by a system of spot checks by officials with a penalty for those who are caught. Sophisticated electronic ticket-issuing machines have now been developed which assist the driver, speed ticket issue, supply a range of statistics and provide more information on the ticket for checking purposes.

SPECIAL CHARGES

The theory of fixing basic charges has already been described, but there are many exceptional rates and fares which are made available in special circumstances.

A railway is expensive to construct but, once established, provision of extra trains within the limits of line capacity is relatively cheap (always assuming that there is not an added artificial complication of paying an independent body for track occupancy as it is used). It has already been noted that if the proportion of overheads within total costs is high, there is a natural desire to secure additional traffic within the available capacity even at a rate that makes only a small contribution to overheads. This is the traffic that needs to be encouraged as the aim must be to spread on-costs over as many units as possible. Occasions arise when the acceptance of a sub-standard agreed charge is commercially justified, for example to maintain goodwill and loyalty. On the shipping side, the Liner Conferences discussed in Chapter 7 are designed to encourage this sort of arrangement. Road hauliers might find it advantageous to utilize at a lower rate spare capacity in a vehicle that would otherwise be wasted.

On the passenger side, special reduced fares in the off-peak periods are an example. Also popular is the offer of special fares aimed at specific groups of people such as students, senior citizens or families accompanied by children. The theory here is that such an arrangement will attract lower income groups who would not otherwise travel. The use of such concessionary fare facilities can always be prohibited on regularly heavily loaded journeys which also helps to exclude those passengers who would have paid the full fare anyway. Another example is for early or late bookings of package holidays and sea cruises where discounts are offered to ensure that patronage is secure or to fill what would otherwise be wasted capacity.

The fares available for air travel are now so complex and bewildering that it is difficult to see how any one principle has been applied other than to maximize revenue from each flight. The application of a global inter-airline tariff as agreed through the International Air Transport Association on behalf of the scheduled airlines as described in Chapter 7 has now been eroded by a proliferation of special fares. They are then supplemented by even lower offers by charter airlines as part of an inclusive package deal. To fill seats, airlines give selective discounts, with conditions and cancellation penalties, in the form of 'advance period excursion' fares (APEX). Some travel agents give even bigger discounts, particularly those who are 'consolidators' – those to whom the airlines dispose of blocks of 'seats' that might otherwise be difficult to sell. But some of the standard fares pay scant regard for distance and there are wide differences in fares even within the same class for the same journey.

Far from IATA's integrated fares structure (which is still in situ), if passenger requirements are flexible (which excludes the essential inelastic demands of travellers who pay the full fare anyway), then, for the best bargains it has become a case of shopping around! Airlines also afford benefits to 'frequent flyers' whereby passengers are given free flights and other privileges in return for past loyalties which might or might not be extended to discount ticket holders.

Another example where fares pay scant regard to mileage is on services where carriers might have exclusive traffic rights. Take, for example, the service from London (England) to Bermuda. The only airline that has a direct link between these two places is British Airways. There is no alternative airline and no alternative mode; a point which seems not to have been overlooked when the tariff was prepared. Bermuda is an island in the Atlantic Ocean some 900 miles short of the North American mainland. But consider the fares. The following table shows the normal published fares (just as an example they are all return fares and based on the higher summer season and valid for more than two months) from London to Bermuda and also to selected points in the USA:

| Place | Fare* | | Mileage from |
	Business class £	Economy class £	London
New York	3147	459	3458
Atlanta	3267	482	4213
Miami	3325	560	4430
Bermuda	2353	916	3444

* Fares as quoted by British Airways

Note in particular the economy class fare to Bermuda and the mileage in relation to the other points where various alternative facilities are available. This suggests that monopoly without control is not a good thing. There are special fares subject to certain conditions which are cheaper and which do include availability on the British Airways direct service. These tickets are generally sold by agents as has been described. But agents also sell tickets valid on other (American) airlines which are circuitously routed via the North American mainland and these can be still cheaper than the 'cheap' tickets on the direct service! Note also the business class fares which exemplify the point, made under the sub-heading 'Class of travel' below, that these passengers, although enjoying better conditions, are paying dearly for the privilege. At least the business class scale pays some regard to mileage.

OVERVIEW

The policies of competition and co-ordination as discussed in Chapter 1 have relevance to and could influence the determination of fares and charges. The economic theory of price fixing in a competitive environment has been considered. Then there is the perfect monopoly which, like perfect competition, is a rare thing. But that apart, in economic terms, the monopolist can either control the supply and let the price fix itself according to demand or fix the price and let demand determine the supply. In other words, he cannot fix both supply and price. In any case, the charges of a transport monopoly would invariably be subject to some form of outside control as will be described in Chapter 7. Nevertheless, the existence of both of these situations in their imperfect forms does leave its mark on the determination of fares and charges.

Carriers operating within both the co-ordinated and the competitive systems will seek to maximize their traffics and hence their revenue. It is likely, therefore, that both will offer special facilities and concessions on the lines that have been described. Competition goes one stage further in that it encourages the provision of alternatives (on the popular routes) which in turn provokes price wars. This is likely to force down fares to even lower levels which is what competition is all about. But those cheap fares may not last any longer than might the services and when the weakest have succumbed the one that is left could, in the absence of any control, have another fares revision! Competition does not mean stability.

In a co-ordinated or monopoly style framework, competition of this kind would not happen. Without the fear of traffic being poached, fares can be pitched higher – except that they could be controlled. But that apart and remembering the concept of what the traffic can bear, carriers might do this at their peril. Even when there is no directly parallel alternative, potential passengers still have the options of going somewhere else, going by private car or not going at all. This is certainly the case where the demand is elastic.

The message that comes out of all this is loud and clear: fares and charges are a useful tool with which to manipulate demand. A system of cheap fares can attract traffic away from one mode and on to another, it can stimulate new traffic and it can discourage existing traffic. Operators can, therefore, within limits manipulate loading patterns to their advantage. Importantly, fare levels (with statutory backing or financial support) could also be used to help tackle the environmental problems which are now beginning to loom large. It would be one way of diverting traffic on to more environmentally friendly modes. More will be said of this in Chapter 9.

CLASS OF TRAVEL

Comfort and space are important considerations for passengers and requirements vary according to the length of journey and fare levels. Although everybody seeks the highest standards, most will settle for less or, for short trips, put themselves to some minor inconvenience rather than meet a price differential. For example, intending passengers at a bus stop will all squeeze on to the first bus rather than wait a minute or so for an emptier one behind even if they know it is there. Nevertheless, there is a minority that will opt for superior accommodation which gives rise to the provision of more than one class.

The availability of separate classes is a controversial issue. It is heard said, and there is some truth in it, that allocating more lavish space for the upper classes can be done only at the expense of the amenities that could otherwise be provided for the majority. The majority, therefore, resent this arrangement. This was the case at one time on the heavily trafficked crush-loaded urban rail services but the trend over the years has been to have one class only on busy short-distance trains. Operators are sensitive on the matter and pay lip service to passenger resentment by disguising the lower class(es) with other names such as 'standard' or 'tourist'. Cruise liners are now often one class only with varying charges levied according to the grade of cabin occupied, which is reasonable enough. One effort to disguise is in China where the People's Republic (where everybody is equal!) does not like reference to first class. Chinese Railways, therefore, advertise their facilities as 'soft class' and 'hard class'. The railways of India at the time of British rule (which ceased some 50 years ago) had four classes which was subsequently reduced to the existing two.

Whatever might be the objections to class distinction there are circumstances where few could deny that separate provision is necessary. In many of the developing countries the accommodation that is provided for the masses is so basic (albeit cheap) and so overcrowded that unless a separate upper class is available, some of the local people and most of the foreign visitors would not be prepared to travel.

An alternative arrangement is the provision of differing quality vehicles/vessels and services with correspondingly different fares. Examples of this practice are not hard to find. For instance, again in China but this time in Taiwan, the Taiwan Railway Administration has numerous different classes with their respective fare differentials but based on the quality and speed of the train. Any one train is therefore composed basically of only one class. Passengers are still paying to a variable scale according to the quality of the service but a by-product of this system is that it avoids the aggravation of seeing passengers in the next compartment or the next coach travelling in greater style. Most of the charter aircraft in the Western world which cater for the masses with their cheaper fares provide another similar example.

In this vexed question of class, it is the airlines that have really gone for segregation. As a standard practice they have basically three classes (some are now going to four) but, like the other modes, have attempted to camouflage any conceived stigma on the lower class by different designations. The second class has a range of exotic titles which vary by airline whilst the third class, being the one used by the masses, is designated 'economy'. Priorities and standards of space, comfort and food are all higher in the upper classes which is expected and is not unreasonable. But on the scheduled airlines it is all on the same aircraft and hence is accomplished in full view of the economy passengers. On landing, the upper classes disembark first whilst the majority sit, watch and wait. The atmosphere suggests an expectation to stand to attention while the exalted minority go on their way! Although the upper classes pay dearly for the privilege, the airlines seem anxious to turn the great bulk of their customers into second-class citizens and people understandably resent this. With arrangements of this kind it is small wonder that there is aggravation and a lack of goodwill.

Having said this, the question remains whether the upper-class passengers are really getting a good deal or whether it is just a moneymaking exercise on the part of the airlines. On a flight across the world, the difference in the price of the cheapest and the most expensive ticket held by any two passengers for the same journey on the same plane could amount to several thousands of pounds. Whatever the upper classes are getting they are paying for dearly, albeit perhaps through a generous expense account. They pay because they cannot face the cramped hurly-burly to the rear of the plane. More comfort and prestige – yes; but at a price and value for money – doubtful. On the other hand, for the airlines it is good economics. Unlike the airports which benefit from large numbers of people, it is the upper-class traffic that makes an important contribution to airline revenue. It helps to subsidize and hence make more attractive the economy class fares even though the point may be missed at the time by the economy-class passengers.

When it comes to class segregation, it is acknowledged that it is a desirable and at times essential facility. But to the passenger it can easily bring with it resentment and ill-will. This chapter is about consumer affairs. It is written from the passengers' point of view and something, somewhere, is rather less than satisfactory.

SERVICE PROVISION

If a trading organization is to have customers its product must be right, the quality must be acceptable, the price must be competitive and there must be a public awareness of its existence. Remember that the movement of goods is pre-arranged. The detailed collections and deliveries and hence vehicle provision and routeings are adjusted in accordance with day to day requirements. But the bulk of the passenger work consists of pre-scheduled services which will run regardless of the actual demand at the time for people to use maybe on an ad hoc basis as the fancy takes them. It is the passenger services, therefore, where the pre-planning of a network has the greatest significance and which was discussed in Chapter 1.

Within an integrated system a transport undertaking would likely have an obligation imposed upon it to provide comprehensive facilities within defined boundaries adequate to meet overall requirements. This could apply particularly to passenger transport within an urban conurbation. It is under these circumstances where standards must be set and 'adequacy' determined that the criteria for service provision has application.

Network planning is also important in an environmental context as there must be good public transport alternatives if people are to be coerced out of their cars. Running services on a purely commercial basis is just not enough for this purpose. The subject of service provision is therefore considered here in the scenario of overall responsibility to provide adequate facilities within a defined area.

From a passenger point of view, the basic requirements of an urban network, which involves any or all of the appropriate modes, may be summarized as follows:

- The walk to and from the nearest bus stop or station at the points of origin and destination should be reasonable.
- Travel between any two points should be feasible with acceptable interchange arrangements if there is no through service.
- Length of wait at bus stops or interchange points must not be excessive and ideally there should be shelter from the weather.
- Fares should be attractively low.
- The service must be reliable and run in accordance with the advertised times and at a reasonable speed.
- The ride should be pleasant and the seats comfortable with little or no standing, particularly on the buses.
- Service information must be freely available and the system must not be confusing (which means a full public awareness of what is provided).
- Services should be available on a daily basis from early morning to late night.

These are the fundamentals and represent the standards as they are likely to be seen by the users. The actual planning function was considered in Chapter 1. To the operator, the above-mentioned requirements are demanding. However, selective prohibitions on the use of private cars is an option that sometime in the future cannot be ruled out and it would be unreasonable to enforce any such restriction if the above criteria was not met. This aspect is considered further in Chapter 9. The requirements listed as needing to be 'reasonable', 'comprehensive' or 'cheap' have not been quantified and in this basic review of the principles, this is not intended. Nevertheless, levels would have to be determined. But they must be acceptable to the users and public consultation is mentioned under the sub-heading 'User Representation' below.

SELLING THE SERVICE

The sale of transport can be promoted by publicity, personal representation and advertising agencies.

Personal representatives are employed mainly in connection with the movement of goods. By regular calls on forwarding agents and individual customers they can, as well as selling the service, advise regarding route-ings, give sympathetic hearings to complaints and quote charges. With passenger travel, such canvassing as is done usually takes the form of calls on travel agents and organizers of party outings who may require bulk transport for sport fixtures, visits, etc.

Advertising can be met with either indirect or direct publicity. Indirect publicity is provided by the promotion of places to visit with an added reference on how to get there. This may need no more than the co-oper-ation of local tourist authorities. Otherwise, it is for the undertaking itself or, if the network is subsidized, by whoever is providing the mone-tary support – such as a local authority – to produce the publicity. Whoever does it, the important thing is that there must be a public aware-ness of what is available. A good map is an ideal way of indicating what services are provided and if drawn to an approximate scale and incorpo-rating other more general features, is likely to be in popular demand. Having reached the people, it then conveys for all to see the existence and details of the public transport system. An indication of the fares should also be included with any special offers highlighted. Equally desirable except possibly where services run at frequent intervals is the issue of a timetable as potential users must be able to ascertain the times of opera-tion. Once people have been attracted they must be retained but this comes back to the quality issues of punctuality, reliability, etc.

Tickets for short and medium distance journeys by bus and tram can usually be obtained on the vehicle. If not, and certainly it is operationally advantageous if they are not, then purchase must be easy at some adjacent point. Potential passengers must not be deterred because there is no read-ily available means whereby they can buy a ticket. For prepaid tickets, therefore, special facilities must be provided and strategically sited sales points in every neighbourhood unit are necessary. For local journeys, although not ideal, this could take the form of roadside ticket machines, but the cost of running manned outlets scattered wherever an occasional demand might arise would be prohibitive. However, as no specialized skill is required for the sale of these tickets, any local shop is suitable and newsagents are the favourite outlets. For longer and less frequently made journeys it is the travel agent that fills the gap, being an agency working on behalf of all carriers, maybe of any mode. In return for this service, a commission is granted on the value of the tickets sold. To the customers there is an advantage here as they benefit from the fact that information and booking facilities are available for everything that is on offer regard-less of the different operators involved.

As the travel agent offers a comprehensive service to passengers, so the forwarding agent offers to manufacturers and traders a service for the dispatch of merchandise regardless of destination, mode of transport or individual carrier. The principle is the same. The consignor has the benefit of dealing with one party, the carrier has ready-made traffic and the agent is remunerated by commission from the charges. An example where the use of an agency is useful is in road haulage where facilities are provided by a large number of small units with few offering a nationwide service. Another example is for international freight which could involve different carriers as well as complex documentation and customs arrangements.

USER REPRESENTATION

User representation in this context refers to machinery introduced by government to protect the interests of the users or to enable them to make their voices heard in instances of dissatisfaction or dispute with the carriers. Originally, the idea was to act as an independent passengers' watchdog when carriers operated in near monopoly situations, such as travel by train or the conveyance of mail.

Briefly, the system of a users' council is that in the event of a passenger or customer having reason to complain about a service or the loss of a facility (such as the closure of a railway station or a post office) there is some third party with overriding power to deliberate. A typical procedure is that if a passenger or customer fails to obtain satisfaction from the carrier in respect of a complaint (and in a monopoly situation the effect of competition would be lacking) then that person can then approach the relevant user body. That body, in whatever form it might be constituted, will then consider the matter and discuss with the carrier concerned. If the complaint is considered justified, and particularly if it involves a reduction or withdrawal of a facility, a recommendation can be made to government. It is then in the hands of the appropriate government minister to consider whatever action might be necessary.

Another example of action in respect of customer interests is for organizations (maybe with official 'encouragement') to make known publicly their policies, standards and commitments. In Great Britain, for example, these statements, which are prepared by the larger utility undertakings, are known as 'Passengers' Charters' and some promise, under certain circumstances, compensation if their declared standards are not met.

A related issue on this topic is public consultation. This is something that has been adopted across the Western world, mainly by national and local government departments and by operators of passenger services within urban areas. It arises when new plans are prepared in respect of such things as land development, proposals for recreational and other facilities, major changes to public transport services and changes of mode, like the provision of light rail transit, etc. The proposals are circulated, the people generally are asked for their views and public meetings might be held. The theory is that the people will then get their say in respect of what they want.

The principle of this sounds good and in many areas it has become a firmly established procedure before any major schemes are embarked upon. However, it is submitted that in practice the system can be something rather less than ideal. Consider the following:

- The people (the electorate) may be too readily influenced by and will react favourably to glossy brochures.
- Only a small percentage of the people usually respond and then often those who are strong in voice but weak in knowledge.
- Individuals look only at their personal situation to see how it benefits their own individual arrangements.
- It is unlikely that there will be any appreciation of the costs involved even though broad estimates might be quoted in the publicity.
- Once the scheme is in place, the preconceived views on what was wanted and what is now there is not necessarily the same.

The danger with public consultation is that the professionals responsible for planning, etc are likely to have a better conception of what the people really want than the people have themselves. They will be looking at the overall benefits as they apply to the area as a whole, the financial implications, the extent to which they will be used and value for money. It is unlikely that the average layman without any technical knowledge of these niceties (even if he thinks he has) will make any effort to appreciate these things.

SEVEN

Regulation and Control

INTRODUCTION

Adequate transport facilities are essential to economic development, the life of the community and the nation generally, and it is the responsibility of governments to ensure that proper provision is made and suitable steps taken to achieve an appropriate and acceptable system. But political doctrines vary and are likely to differ between nations and between different governments of the same nation in the event of change. It follows that both the objectives and the methods of achieving those objectives also vary. Hence there is no common worldwide statutory background to transport provision and operation although, as has already been said, now, at the close of the twentieth century, one thing which is virtually worldwide is a swing towards privatization. Whilst certain aspects of government control are (or should be) inescapable such as, for example, the provision of a national highway network and the determination and enforcement of adequate safety standards, other involvements are likely to be dictated more by pure politics. The attractiveness of transport to the vote-seeking politician has already been noted. The very fact that transport is essential for the sustenance of everyday life and that its services are forever in the public eye make it a prime target for election manifestos. Many of the issues are subjective in character and in any case there is always more than one way of achieving an objective. Furthermore, government policy, even when not specific to transport, can materially influence transport management. Employment legislation is an example; care for the disabled and the protection of the environment are others.

The structure of the transport industry as such was considered in Chapter 4 where the fundamentals of public and private ownership were noted. The implication here is that it can condition the extent of control as the terms of reference of the public sector undertakings are more likely to be explicit and embodied in the statutes. The existence of either of these sectors is determined by the political persuasion of national governments or by legacies from their predecessors and the extent to which public transport is regarded as a social service. Where the latter situation prevails and a subsidy in some form is applied, this did, over the years, encourage public ownership as the principle of propping up private enterprise out of the public purse was at one time regarded as an anathema. Now, with privatization fashionable, machinery has been found to maintain at public expense non-commercial but socially necessary services which have entered the private sector. In fact, as was seen in Chapter 4, there are now instances where private capital has been used to support public transport. But be quite clear: privatization does not also mean deregulation and in no way need privatization inhibit controls even though it has not been unknown for the two features to be introduced at the same time. Buses in Great Britain are an example.

Statutory requirements regarding safety are (or should be) imposed on all operators who may also be subject to commercial controls. Inevitably the detail will vary country by country and in this study of the principles it would be impracticable and unnecessary to attempt to catalogue the mass of different regulations and legal requirements which apply to all of the carriers in all of the modes worldwide. What must be considered, however, are the commercial issues that might be regulated and this could be influenced according to whether the services are operating in a competitive or a controlled monopoly situation. If competition is to be restricted then obligations need to be established in order to compensate for the loss of the effects of the economic forces of supply and demand.

There are also voluntary forms of control through internationally agreed standards of operational conduct, etc which are established through international bodies. All of these issues will be considered.

FINANCE

Reference was made in Chapter 1 to the essential nature of transport and although many of the passenger services are now no longer profitable, there would be disruption and hardship if they were not maintained. For example, most railway systems and urban rapid transit networks would close and large tracts of country would be without buses. The results would be untenable and are not an option. If the facilities that are deemed to be socially necessary or desirable cannot be provided commercially, then it is the responsibility of government to do something about it. The only variable factor then is how much money should be involved. There are various ways of allocating revenue support but whatever method is adopted, if a transport undertaking is subsidized by a third party, then whoever provides that support will expect some say in the services that are provided. That is why the subject of finance is touched upon in this chapter.

The various methods of subsidizing public transport include:

- government writing off an annual deficit;
- government stipulating its required network and then meeting the deficit thereby incurred with the operator working to a 'public service obligation' requirement;
- government stipulating what services or journeys it requires over and above what can be provided commercially and maybe then putting its requirements out to tender.

Examples of the first option may be found in some of the developing countries, particularly where revenue leakage produces heavy deficits and where Western aid is expected. The second alternative is a requirement often placed on national railway systems. It is an appropriate method as the railway is an essential albeit unprofitable national asset. Some countries also support their national airlines in this way, perhaps for traffic reasons but sometimes also for prestige. Air France, for example, is still supported from public funds. The third possibility is more appropriate for buses where several operators could be in a position to provide a service and are able to offer competitive tenders. This alternative has merit in that it can instil an element of competition and thereby keep down costs in what is still a regulated environment. More is said below on this particular option.

The point that is relevant to this chapter, however, is that whoever puts up the money has a say in the services that will be provided.

REGULATORY CONTROL

Quantity control

Control of service provision (where it is applied and laid down by statute) is likely to be administered through specialist government-appointed bodies or individuals. Quantity control means that operators are not free to provide all of the services they might wish. Before alterations are made, approval would be needed and the independent authority would be required to listen to objections from other operators who might consider that their interests would be jeopardized and also from users. Such a system protects established operators, ensures a degree of co-ordination and avoids wasteful competition but it does, of course, become superfluous in the extreme case of a statutory monopoly. Here it is necessary to ensure only that the operator does not exploit that position to the detriment of the services that are offered. The independent authority would then concern itself with overall provision, complaints about lack of service and requests for fare increases.

Quantity licensing, therefore, controls services and administration through an independent licensing authority is most appropriate particularly where a number of different operators are involved. This means that the arrangement is suitable for road transport and can also be applied to air transport. The mechanics of a typical system could be as follows:

- establish a licensing authority with discretionary powers to approve the introduction or variation of services and fares;
- operators apply to the licensing authority for permission to introduce or vary a service;
- licensing authority publishes applications from operators for all interested parties to see;
- interested parties object to such applications should they see a reason so to do;
- licensing authority considers applications and listens to evidence from each side, conducting a public hearing if necessary;
- licensing authority gives a decision which may be an approval, a refusal or a modified grant;
- decisions subject to right of appeal by either side.

The 'licensing authority' would be under the chairmanship of a government-appointed officer acting in accordance with prevailing legislation and declared policy. Aspects of a service for which specific approval might be required could cover some or all of the following:

- route
- stopping places
- timetable
- fares
- vehicle types

Although it has been mentioned that this type of control could be applied to air transport as well as to buses, as there is here a strong international element it does mean that there would need to be bilateral agreements at government level before the domestic issues could proceed. Subject to some variation, the system could also be applied to road haulage and particularly where, for example, there might be a national policy to protect the railway.

All of the five aspects as listed in the previous paragraph could be stipulated when services are put out to tender, the system of which was referred to in Chapter 5. It had relevance there in that it produced yet another master whose directions and requirements had to be met by the operator. It has relevance here as it provides a positive form of quantity control administered through an authority charged with a responsibility to provide a co-ordinated network of services. If an operator is granted a franchise to work any particular route or to serve any specific territory then he must abide by the terms imposed. Acceptance by the authority of a tender will likely be based on the most effective and economic application of funds and at this stage, operators will be in competition with each other. Once the service(s) become operational, however, there is control. Contracts can be let for specified periods either on a fixed price or minimum subsidy basis with the operator either taking the risk or on a revenue-guaranteed basis.

Quality control

Even in a deregulated society there must still be a measure of control and it is the quality licence that should ensure that operators observe adequate safety standards. Matters which might be covered by a quality licence are:

- qualifications of staff
- adequate financial backing
- limitation of working hours worked by certain staff
- standard of maintenance of vehicles and equipment
- suitability of vehicles, vessels or aircraft

Generalized regulations involving safety need to be applied in any case to each vehicle, ship or aircraft, to the permanent way of a fixed track system and to the navigation arrangements of ships and aircraft. Traffic on roads must be regulated with adequate safety standards declared and enforced. It is only the quantity control of public transport that may be left to discretion. There should be adequate quality control with or without the imposition of quantity control.

TRADE ORGANIZATIONS

Preamble

The general term 'trade organization' is used here to refer to the different associations involved with the provision of transport which have grown up over the years. Whilst not exercising control in the literal sense, some prepare standards, particularly in respect of safety and environmental matters, maybe with recognition by national governments and with which the more reputable carriers set themselves out to comply. They could also be the medium through which inter-government agreements might be implemented. It is appropriate, therefore, that they should be considered in this chapter. Associations that act in the interests of the transport workforce, ie, the staff associations and the trade unions, are not for the purpose of this chapter regarded as a part of this classification.

A trade association is a body that acts in the collective interests of the many different undertakings that function within a particular industry. Such an organization might act on behalf of its members by presenting, for example, a united front on representations to parliament when changes to legislation are contemplated. It might also provide a service that meets a need of members such as the dissemination of information on technical or legal matters, guidelines on costing and charging, or the preparation of publicity for the industry with a view to the procurement of a good public image. Other organizations within this broad classification do not necessarily have a membership as such but are agencies of or are recognized by the United Nations. They prepare codes of conduct and safety rules and more recently advise in connection with the preservation of the environment.

All of the organizations to be described here are international in character. It may be noted, however, that with some of those mentioned there are also parallel national bodies which themselves are likely to be members of their international counterparts and which offer a similar service within their own domestic confines. However, a brief review of a selection of the most appropriate international bodies is sufficient for a general study of the principles of transport and this now follows.

International Road Transport Union (IRU)

The economic recovery of Europe that followed the Second World War resulted in a steady growth of trade with international transport playing an important role in fostering closer relationships between the different countries. It was under these circumstances that an international organization to bring together the various national associations concerned with road transport became necessary and it was for this reason that the International Road Transport Union was formed in Geneva in 1948. It was subsequently granted consultative status both by the United Nations Organization and by the Council of Europe. It is represented in Brussels (Belgium) by a Permanent Delegation and Liaison Committees to the European Union; matters pertaining to central and eastern Europe are dealt with within a Commission grouping the Associations of all countries in this region. The activities of the Union embrace road haulage (including both hire and reward and own account operations) and passenger transport by coach and taxi. Membership (156 active and associate members) has now extended beyond Europe to a total of 64 countries in all of the 5 continents.

The basic objective of IRU is to promote the development of international road transport in the interests of carriers and the economy as a whole. It co-ordinates viewpoints, brings together what are sometimes divergent conceptions of transport and aims at improving the conditions of its members by rationalizing their work and encouraging the adoption of the latest techniques. Groups of experts make their contribution in specialized fields such as combined transport, customs and legal problems, environmental protection and road infrastructure.

One of the main concerns of IRU is the reduction of frontier controls and to this end it administers the issue and use of TIR (Transport International Routier) carnets which simplify customs formalities. The TIR carnet is an international customs document that provides free movement of goods carried in vehicles or containers in transit across frontiers that lie *en route* between the points of origin and destination. The possession of a carnet exempts the carrier from customs examination of goods at all frontiers except the first and the last on any particular journey and in normal circumstances the payment of duty at intermediate customs points is not required. TIR vehicles or containers have the further advantage that, with certain exceptions, they may be cleared at inland centres rather than at ports where delays are more likely to be encountered. Vehicles used in connection with this arrangement must be of an approved design with their goods-carrying areas closed and capable of being sealed and the carriers concerned must be on a specially approved list. Satisfactory operation of the TIR carnet system does of course require the co-operation of the customs authorities and its use by only reputable carriers. IRU works to ensure that goodwill is preserved between all parties in this respect.

Other achievements of IRU include the establishment of an international mutual aid service for goods and passenger operators and the implementation of an international star classification system for touring coaches. It also encourages professional training through the organizing of seminars aimed particularly at the central and eastern European countries. Of current importance is the Union's consciousness of the need to preserve a satisfactory environment. To this end it has drawn up a charter for sustainable development whilst bearing in mind the interests of its members. In so doing it recognizes the importance of the application of technological progress to vehicles, infrastructure and logistics and again also to professional training, all of which can reduce pollution at source, save fuel consumption and improve road safety.

International Road Federation (IRF)

The International Road Federation is a non-profit service organization established in 1948. Its purpose is to encourage better road and transport infrastructure worldwide and to help in the application of technology and management practices aimed to give maximum economic and social return from national road investments. IRF has over 600 members – mainly from the private sector – in over 100 countries. Its offices are in Washington (USA) and Geneva (Switzerland).

In the USA, through fellowships and executive conferences, professional study programmes and a video training library, IRF supports advanced training programmes for highway and traffic engineers from around the world. In Europe it lobbies the EU for improved road infrastructure investment, it organizes road-related conferences in developing countries and provides road statistics and data banks, technical and economic data and other educational material.

IRF co-operates with international organizations with similar objectives and has consultative status with UN organizations, the Organization of American States (OAS), the Organization for Economic Cooperation and Development (OECD) and the Council of Europe. To further the exchange of knowledge, it organizes world meetings every four years and regional conferences in between.

International Union (Association) of Public Transport (UITP)

The organization now known as the International Union (Association) of Public Transport was founded in Brussels (Belgium) in 1885 as the Union Permanente de Tramways and is dedicated to urban mobility. Membership is open to public and private sector mass transit operators, national mass transit companies and associations, decision-makers and transport policy-makers, institutes of transport studies and research and transport suppliers and manufacturers. Membership is drawn from 70 countries throughout the world.

The objectives of UITP are to study all questions on public transport to find progressive solutions to problems that affect the mass transit sector. It also represents and serves the interests of its members through direct contact with them and with other international organizations. It has been granted consultancy status by the United Nations. UITP does to some extent share common themes and has similar purposes to those of the Transportation Research Board (TRB) which is based in the United States of America and has primarily a North American constituency. Both organizations stimulate, correlate and disseminate the results of research and undertake special studies. However, whilst TRB is multi-modal and serves as a clearing house for transport research in all modes, UITP (which, unlike TRB, is a worldwide organization) has its focus on mass transit operators along with a strong interest in all aspects of urban mobility and the research needed to achieve it. To deal with problems that are of a particularly complex nature and demanding continuous research, a number of international study groups have been appointed within UITP. Activities covered by these committees and working parties are as follows:

Activities related to all modes of transport
- transport economics
- traffic and urban planning
- regional transport (rail and road)
- organization and data processing
- human resources
- research and development
- standardization
- industry liaison

Also regional committees
- European Union Committee (in cooperation with national associations)
- Asia/Pacific Committee

Also activities related to one mode of transport
- metropolitan railways (operations, rolling stock, fixed installations, finance and commerce, electric installations and safety systems)
- commuter and regional railways
- light rail
- buses (design and maintenance, operations)
- waterborne transport

It will be recalled that IRU also has interests in road passenger transport. Whilst, however, IRU caters for the interests of only the road transport industry but, as a whole, embracing own account and hire and reward road haulage and the longer distance and touring coaches, UITP caters for more than one mode but within the one particular sector, namely urban and regional passenger transport.

International Union of Railways (UIC)

Railways have been in existence for about 170 years and over this period various bodies had dealt with specific European international problems. But the need arose for a single overall organization that could handle all questions. In the event and in the aftermath of the First World War, a series of intergovernmental conferences were held. These culminated in the Conference of Paris in 1922 when the International Union of Railways was founded with its headquarters in Paris and with chairmen drawn originally from the French railways.

One of its first tasks was to simplify through running between neighbouring countries. During the Second World War activities virtually ceased but after that the process of co-ordination was further developed. UIC now has a total membership of 137 railway undertakings spread over all of the continents of the world; but it includes 31 affiliate members which are entities conducting activities complementary to those of the main line railways.

The role of UIC is to promote co-operation between members at world level and to carry out activities designed to develop international rail transport. It maintains and develops the overall coherence of the railway system throughout Europe and builds the framework for interoperability to improve railway competitiveness. To achieve this goal it strives to encourage state of the art technology and modern management methods among its members. It also prepares statements and common position papers to promote the role of rail transport and represents the rail sector with observer status at the United Nations.

The tasks of UIC include:

- preparation of standards, regulations and recommendations to facilitate international traffic;
- working with outside bodies representing the railways;
- development and management of international projects in the field of passenger and goods transport, infrastructure management and research;
- promoting exchanges of information and experience;
- concluding agreements with other international bodies, ensuring co-ordination and consistency on rail matters.

Of particular interest is the development of a high-speed rail network in Europe (for this purpose a 'high-speed train' is generally regarded as one which travels in excess of 125mph (200kph)). High speed rail projects were first realized in the 1960s with the Shikansen trains in Japan which succeeded in almost doubling the passenger traffic. The principle then spread to routes in Europe and by the 1990s France in particular and to a lesser extent Germany, Spain and Italy had routes worked with high-speed trains, again with encouraging load factors. But they were all national services and it was clear that if their potential was to be fully developed the network would need to be integrated on an international level. Accordingly, UIC, in conjunction with the Community of

European Railways (a body which represents the railways of the European Union member states plus Switzerland and Norway when in dialogue with the Community institutions), commissioned a study and proposals were prepared.

A pan-European network has been produced which includes both new lines with speeds around 160mph (256kph) and upgraded lines with speeds around 125mph (200kph) plus a backup of a series of interconnecting links to serve all of the major European commercial centres. Based on a 2010 time horizon, estimates suggest that with such a network, for journeys over 50 miles (80km), railway traffic would likely increase by 73 per cent, 40 per cent of which would come from the roads, 31 per cent from airlines and 29 per cent being newly generated. With trains being city centre to city centre it is considered that they would have a strong advantage over air transport for journeys of up to about 500 miles (800km). The scheme received a favourable reception at the EC Commission and is now in the course of implementation. High-speed international trains in Europe are now a reality and the current core of the network is the 'Eurostar' services which link Paris and Brussels with London and the 'Thalys' services which link Paris with several destinations in Belgium including Brussels, Amsterdam and Cologne.

A particular complication in producing the high-speed network to which the Task Force is committed is the historical disparities of the different systems in terms of track gauge, electrical power supply and signalling systems. In Chapter 3 the importance of standardization was stressed. Fortunately, most of the European systems were built to the standard track gauge (4ft 8½in/1435mm) except for Ireland, Spain, Portugal and Russia where the results of the choices at the time have now come home to roost. It is significant that Spain has since elected to build its first high-speed line (between Madrid and Seville) to the (narrower) standard gauge. It is likely that the different electrical energy supply systems will just have to be lived with but UIC is co-operating with the EU and the European supply industry in developing a standard control system and also a unified radio communications system. These achievements will improve the European dimension of rail operation, at first for high-speed trains and then for conventional passenger and goods trains as well.

International Maritime Organization (IMO)

The International Maritime Organization was established in 1959 as a specialized agency of the United Nations and the only one concerned solely with maritime affairs. Membership embraces 156 countries. Its interest lies mainly in ships used on international routes and its objectives are to facilitate co-operation among governments on technical matters affecting shipping with special responsibility for the safety of life at sea and the prevention of marine pollution.

In order to achieve its objectives IMO has promoted some 40 international conventions, protocols and other instruments concerning maritime safety, the prevention of pollution and other related matters. The initial work on a convention is normally accomplished through committees which are open to participation by all member governments. These committees represent:

Maritime safety: dealing with
- safety of navigation
- radio communications and search and rescue
- bulk liquids and gases
- fire protection
- standards of training and watchkeeping
- carriage of dangerous goods, solid cargoes and containers
- ship design and equipment
- stability and load lines and fishing vessels safety
- flag state implementation

Marine environment protection: dealing with pollution prevention and control
Legal: dealing with legal matters following accidents, etc
Technical co-operation: dealing with co-ordination of technical assistance
Facilitation: dealing with facilitation of international traffic (formalities, documentation, etc)

The resultant draft instruments produced by these committees are submitted to a conference to which delegations from all States within the United Nations system are invited, regardless of whether they are members of IMO. It is the conference which adopts a final text which is submitted to governments for ratification and implementation. The requirements of such a convention become mandatory on those countries which are a party to it. Other codes and recommendations adopted by IMO which are not in themselves binding on governments are generally implemented through incorporation into domestic legislation.

Of particular relevance in this context is the prevention of marine pollution. The most spectacular instances of environmental damage are, of course, those which arise from accidents to oil tankers of which the wrecks of the *Torrey Canyon*, the *Amoco Cadiz* and the *Exxon Valdez* are examples. As far as oil is concerned, however, the most common pollution incidents occur during terminal operations when oil is being loaded and discharged and also as a result of normal tanker operations involving the cleaning of residues whilst the ship is returning empty. Furthermore, all types of vessels can be responsible for pollution when being cleaned of bilge and fuel oil during dry docking. But it is not only oil that is a pollutant factor. Garbage and sewage from ships have over the years been dumped into the sea. At one time most of the material was assimilated by the marine environment without seriously harmful effects but now, with the increasing scale of operations together with the widespread use of plastics (which are non-biodegradable and hence stay in the sea for many years) and radioactive wastes, it is a different matter.

In 1973 IMO adopted the International Convention for the Prevention of Pollution from Ships; this was modified in 1978 and is now known as MARPOL 73/78. Since that date there have been further annexes and together they cover such measures as the construction of tankers and methods of crude oil washing, the carriage of noxious substances, sewage and garbage disposal (particularly plastics), dumping at sea and air pollution from ships' exhaust emissions. The measures introduced by IMO since it was founded have provided a framework for reducing marine pollution from ships and, as a result, the problem has been eased. However, the fact that it still exists indicates that although the measures are there, more enforcement is necessary.

An interesting case study lies in the garbage disposal problems being experienced in the Caribbean. The threat there is serious as the area is a magnet for cruise shipping which is growing in popularity: cruise liners with a full complement of passengers generate as much garbage as a small town. A ship should, of course, be able to dispose of its rubbish when in port and Annex V of MARPOL deals with this aspect. But as yet Annex V has not been ratified by all countries and this includes many of the small island states of the Caribbean which do not have adequate resources for this purpose. Under these circumstances it is tempting for ships' captains to dispose of garbage illegally at sea. As a result, rubbish will in due course litter the beaches and, ironically, damage the very environment which attracted the tourists in the first place. The Peninsular and Oriental Steam Navigation Company (P&O), for example, which has an extensive cruise programme in the Caribbean, is on record as recognizing that MARPOL is the background against which any operator should work and has publicly declared itself committed to meeting its requirements. It has adopted such measures as the utilization of sophisticated waste-handling equipment and incineration plant, segregation of waste, elimination of plastic containers for food and toiletries, elimination of environmentally unfriendly chemicals and paints, and employment of fuel efficient propulsion systems.

The work of IMO often runs parallel with the activities of other agencies in the United Nations family (including ICAO which is discussed below) in connection with safety at sea and in the air and close contact is maintained with them on projects of mutual concern. There are formal agreements with the United Nations, the International Labour Organization (which has a particular interest in employment conditions everywhere), the Food and Agriculture Organization, the Atomic Energy Agency and many other bodies within the United Nations system. The latter's work programme covers such subjects of interest to IMO as the application of atomic energy to ship propulsion, the disposal of radioactive waste from such ships and the carriage of radioactive cargoes.

In consultative status with IMO are many non-governmental international organizations representing shipowners, trade unions and other interests and bodies concerned primarily with shipping matters, especially from the safety aspect. These all play a valuable part in assisting IMO with its work.

International Chamber of Shipping (ICS)

Formed originally in 1921 but assuming its present title in 1948, the International Chamber of Shipping (ICS) is an association of 42 national organizations worldwide which together represents more than half of the world's merchant tonnage. The principal role of ICS is to promote the interests of its members by providing a forum for the discussion of matters of mutual concern, by seeking to co-ordinate members' views and by representing those views internationally. The activities of ICS embrace not only general policy but also technical, legal, and safety issues as well as the facilitation of trade procedures and documentary matters. ICS has consultancy status with a number of inter-governmental bodies, notably with IMO referred to above.

ICS shares a secretariat in London (England) with the International Shipping Federation (ISF) which co-ordinates the interests of maritime employers. ICS maintains close links with a number of other industry organizations and produces industry guides on various aspects of ship safety, prevention of pollution and trade procedures.

Lloyd's Register (LR)

Lloyd's Register is the world's premier ship classification society and a leading independent technical, inspection, certification and advisory organization. Its ancestry can be traced to Edward Lloyd of the City of London in whose coffee-house it had its birthplace in 1760.

From the time it was founded, LR has been engaged in the classification of ships and throughout those years it has recorded the results of its work in a series of Register Books. It is the classification that sets and maintains standards of quality and reliability for all types of vessels. It establishes requirements for the design and construction of ships and their essential machinery and maintains these standards by periodic surveys. In many countries classification is required for ship registration and is usually a prerequisite to obtain the statutory certification required for international voyages. Over 104 million gross tonnage of shipping representing over 20 per cent of the world's merchant fleet is classed by LR in accordance with its own rules for the construction and survey of ships. Although independent of any government or other body, more than 135 national administrations delegate LR to survey ships to ensure that they meet national and international safety legislation.

But LR is not concerned solely with shipping. It also has an industrial division offering inspection services for land-based industrial projects, an engineering division offering technical advisory services for the marine, offshore and industrial sectors and an offshore division providing classification, verification, technical advisory and inspection services for the offshore sector.

Direction of LR's affairs is in the hands of a General Committee representing the main interests of the international maritime, petroleum and industrial communities and a Technical Committee consisting largely of representatives nominated by marine and industrial institutions. A Management Committee provides the necessary impetus and management control.

Baltic Exchange

The Baltic Exchange provides a marketplace where shipowners worldwide who are seeking cargoes for their ships (mainly dry bulk carriers and tankers), and merchants and shippers, again from across the world, who are seeking transport for their merchandise are brought together. The Exchange publishes daily its Baltic Freight Index (BFI) which is an indicator of the dry bulk market comprising eleven trading routes. This is the index which forms the basis of freight futures contracts. Additionally the Exchange publishes a Baltic Handy Index (BHI) twice weekly to cover the smaller, geared vessels. Also, the Baltic International Tanker Routes (BITR) is published daily which is an assessment rather than an index of seven international oil routes. Lastly, there is a daily publication of a dry cargo fixture report.

The Exchange was founded in the early eighteenth century in a coffee-house in London (England) when tallow shipped from the Baltic seaboard was the major trade. It has now progressed from a one-time venerable London institution to an international market centred in London with state-of-the-art telecommunication and electronic information transfer although the essential function of the Trading Floor has remained central. It now has a membership of some 670 companies worldwide and London accounts for about 30–40 per cent of the world's dry bulk charters and around 50 per cent of all tanker charters. Freight is still its main activity but now more diverse cargoes dominate such as grain, iron ore, coal, bauxite, phosphate rock, wood, sugar, minerals, fertilizers and agricultural products. The important wet cargoes include crude oil together with its derivatives as well as chemicals and gas. Of growing importance is the charter of container ships. There are other non-market members who do not actively trade in the market but who wish to be associated with it, such as marine lawyers, arbitrators, insurers, financiers, classification societies, etc.

The Exchange does also have other functions. Baltic brokers arrange sales and purchases of vessels and ships for demolition. Also, some traders buy and sell commodities or supply bunkers. An Air Market is another activity whereby airbrokers within the Exchange are able to arrange bulk or specialized transport by the charter of aircraft.

Essentially, the Exchange caters for the charter market whereby shippers hire the complete vessel, etc in bulk for their specific use. The traffic of shippers who wish to dispatch merchandise by the regular sailings of shipping lines is covered by a system of liner conferences which is considered below.

Liner conferences

Liner conferences as a sub-heading is somewhat different in character from what has otherwise been considered in this chapter. It is not a title of any one organization but is more descriptive of a system and within that system; there are many different conferences. As, however, it is the purpose of each liner conference to act in the collective interests of its members, it is appropriate to include reference to the system here.

Where several shipping lines operate in the same trade, it is common practice for them to form an association known as a 'conference'. Basically, a conference is simply a meeting of all the participating lines (of any nationality) that serve a particular route or geographical area for the purpose of evolving a regular pattern of sailings to give the maximum service to shippers and to reach common agreements on rates. According to the importance of the trade, conferences range from informal associations to well-developed organizations with a permanent secretariat. If liner companies are to invest in new ships, keep them on particular routes when on occasions they could be used more effectively elsewhere, maintain rates at a stable level, carry unattractive cargoes and serve difficult ports, then they require some form of protection from outside opportunists who are not prepared to give such a service but seek only the most lucrative traffics.

This is why the voluntary system of liner conferences has evolved. There is generally no legal protection from the 'pirate' ship and the conference must therefore work to a system that attracts and retains regular customers.

It is not, however, only the shipping lines that require some sort of guarantee. Shippers (being the customers) need to have recourse to services that are regular and reliable and they want to be able to sell, knowing that the business which results will find transport regardless of the prosperity of the trade at the time. They also want stability of rates and equality of treatment with their competitors without necessarily being tied to one particular carrier.

For these reasons, conferences have become a part of the liner system. To get some assurance of support, liner conferences offer to regular shippers a special discount on tariffs, provided that they confine their custom to one or other of their members for a defined period. This is known as a 'conference tie' which is in the form of a deferred rebate and, except for general rate cutting, is the only protection that lines within a conference have against outside ships. Shippers are free to deal with whom they please but if they do have business with an outside ship (without permission from the conference) they will lose their special rebates. For this reason, some people tend to see this as an organized cartel.

International Civil Aviation Organization (ICAO)

The International Civil Aviation Organization is the agency of governments that creates world standards for the technical regulation of civil aviation. Its aims and objectives include the development of the principles and techniques of international air navigation and it fosters the planning and development of international air transport. The following brief review of the background which led to the establishment of ICAO will help to clarify the situation.

Cross-border transport produces political complications and this is particularly so in the case of air transport which is the one mode that attempts to provide a unified global network of regular timetabled services. During its relatively short period of development there are, historically speaking, four notable milestones, being:

- The Warsaw Convention, 1929
- The Chicago Conference, 1944
- The Bermuda Agreement, 1946
- The Hague Protocol, 1955

The Warsaw Convention focused on legal issues and the complexities of conditions of carriage which arose from international services, remembering that national laws cannot be enforced in foreign territory. A set of rules and regulations paying particular regard to legal liabilities was therefore produced which was to apply to all participating nations upon ratification by their individual governments. In due course, however, the necessities of the Second World War produced a network of air services run for essential wartime traffics regardless of any political considerations. Matters such as territorial boundaries and agreed charges were of small concern to nations at war and it was realized that many legal and commercial complications would arise if these or similar services were to be maintained on return to peacetime conditions. For this reason, the government of the United States of America convened a (wartime) conference at Chicago in 1944 which was attended by representatives of allied and neutral nations. It was this conference that gave birth to ICAO for the purpose of administering the principles which were enunciated. It is a specialized agency of the United Nations.

The 1944 Convention deliberated on the future of international civil aviation and endeavoured to achieve multilateral answers to the questions regarding who could fly where and at what price. Accordingly it produced two agreements, being:

- an International Air Services Transit Agreement which made provision for the aircraft of any signatory power to fly over or land for non-traffic purposes in the territory of any other signatory power, and
- a more comprehensive International Air Transport Agreement which provided for, among other things, the carriage of traffic between member states. The privileges resulting from this Agreement became known as the 'Five Freedoms', which were:
  1. The right to fly over the air space of any contracting state without landing.
  2. The right to make a technical landing (eg for refuelling) in the territory of any contracting state.
  3. The right to set down traffic in the territory of any contracting state, providing it originated in one's own country.
  4. The right to pick up traffic in the territory of any contracting state, providing it was destined for one's own country.
  5. The right to carry traffic between the territories of any two contracting states.

The first two of these freedoms have been generally adopted. The remainder are still largely being resolved only on a bilateral basis and it was the Bermuda Agreement of 1946 which was the benchmark of bilateral air transport agreements. This agreement was between the USA and the UK and was the first of what is now around 4,000 such agreements which are signed and registered with ICAO.

Review and development of international air transport by ICAO on behalf of governments has been an ongoing process and certain amendments affecting the legal issues were made through the Hague Protocol of 1955. A radically reformed system of compensation is now being developed in conjunction with the airlines (through IATA, see below).

To comply with its terms of reference, ICAO has established the following committees:

- Air Navigation Commission
- Air Transport Committee
- Legal Committee
- Committee on Joint Support of Air Navigation Services
- Finance Committee

The ultimate authority of ICAO is the Assembly at which all member states have the right to be represented. It maintains close liaison with, *inter alia*, IMO, which has already been discussed, and the International Air Transport Association, reference to which follows, and which is represented at many of the ICAO meetings.

International Air Transport Association (IATA)

The International Air Transport Association is the association of the world's scheduled international airline industry. Originating in 1919 as the Air Traffic Association and limited to a European dimension until the erstwhile Pan-American Airline joined in 1939, the Association in its present form was founded in 1945 in Havana, Cuba. It now groups together some 240 airlines, including the world's largest, and together they fly more than 95 per cent of all international scheduled air traffic. The commercial objective of IATA is to ensure that people, freight and mail can move around the vast global airline network as easily as if they were on a single airline in a single country. Its operational objective is to ensure that members' aircraft can operate safely, securely and efficiently under clearly defined and understood rules.

IATA works towards these objectives through representation, by setting standards and by selling a range of services for the travel industry:

Representation: IATA makes representations to achieve a more favourable operating regime for its members, for example endeavouring to secure more airways and airport capacity, fairer taxation policies, recognition of air transport's regard for the environment, etc.
Standardization: Airlines, through IATA, designed the standard passenger ticket and the air cargo waybill, for example. Evolving standards for accounting, data-flow, automated ticketing, baggage checking/trac-

ing, care of disabled passengers, security, elimination of government bureaucracy and aircraft scheduling are all examples of its ongoing work.

Services: IATA is the largest single provider of airline training in commercial subjects. This includes, for example, training in basic travel-agency work, fraud prevention, revenue accounting and airline management, computer programming, etc.

Traffic co-ordination is one of the oldest services offered by IATA. Reference was made to the Chicago Convention under the ICAO subheading where it was noted that the last three of the 'five freedoms' are still for the most part settled by bilateral agreements between the governments of pairs of countries. Whilst it is governments that decide on international traffic rights and fares, IATA is the forum in which they are negotiated and a series of Traffic Conferences have been instituted for this purpose. At one time it was the IATA fares that were always charged, being uniform fares set at a level which avoided wasteful competition. However, with more fares worldwide now being set by market forces, tariff conference proposals have tended to become maximum reference points. Nevertheless, it is the conference fares that enable journeys to be made which involve more than one airline with just one payment and one ticket and it is IATA that divides up the revenue as appropriate ('interlining'). Whilst only about 15 per cent of the total international passenger traffic on IATA airlines is now interlined, it is still substantial as even then it represents about 1 million passengers per week.

To quote other examples of the work of IATA, an important function is that of acting as a clearing house. This enables the mutual debits and credits of members to be set off in total one against the other which thereby minimizes the need for actual cash transfers. Also, as was noted under the ICAO heading, there are workings to reform the system of compensation for injuries, etc within the Warsaw Convention framework. IATA is participating with member airlines in the preparation of revised arrangements. Topical in today's climate is the environment and with it comes supporting legislation worldwide. IATA disseminates such information, monitors international debate and promotes common positions of the industry on environmental issues. These include support through ICAO for a co-ordinated phase-out of old aircraft types and noise-compatible land use planning. IATA is building its own data bank.

In short, whilst ICAO is the regulatory body at government level, in the course of its work it co-operates and consults with the airlines and it is the airlines that are represented by IATA.

EIGHT

Transport and the Environment

PREAMBLE

Throughout this work oblique references have been made to the environment. At the outset it was said how the early lines of communication forged a pattern of development which in turn determined the location of population and industry and how this influenced the surrounding scene. Then, in sparsely populated regions it was transport that enabled incursions into and subsequent opening up of hitherto virgin territory. In more recent times, new towns have emerged and even new capital cities, all of which have again been made possible by the availability of adequate transport facilities. As a result, natural habitats are being destroyed with serious repercussions. There is now a deterioration of what is a changing landscape, the air is being poisoned and the rivers and seas polluted and this is all contributing to a destruction of vegetation and wildlife. Some of these features are just unpleasant. Others are destined to have more drastic and lasting effects as our health is jeopardized and the balance of nature upset. These are the inevitable consequences of increasing human activity and development.

Another feature, and one that is also of major concern, is the consumption of the world's natural resources, some of which are finite. Even so, this is still regarded as a necessary part of everyday life if present-day customs, practices and the overall standards of living that have now been achieved are to be preserved.

What then is the environment? It is simply the condition of our physical surroundings which in turn influence the quality of life. In the course of its provision, transport has had repercussions which have certainly not been negative but if in so doing it has also brought with it unacceptable levels of space occupancy, poisonous effluents, noise, etc, then those issues must be addressed.

The two popular conceptions of environmental damage are pollution and traffic congestion, both of which have a transport connotation. Cars and heavy haulage vehicles are usually seen as the principal offenders on both counts, together with their concomitant demand for roads and more roads. Certainly, pollution must have a strong input to any discussion on the environment but the subject has more to it than that. This chapter will draw together all of the various facets which collectively constitute the environment and examine to what extent transport is responsible for its deterioration. This is a necessary prelude to Chapter 9 which will suggest possible courses of action that might perhaps ameliorate the otherwise devastating effects that must come.

Under a broad umbrella of 'environmental issues' the following detrimental features can be identified:

- lack of space to absorb population growth which although an environmental problem is a rapidly approaching physical problem as well;
- consumption and exhaustion of the earth's natural resources;
- scarring of the landscape, being visual intrusion;
- air and water pollution;
- noise and vibration.

However, although the loss of rainforests, wildlife habitats, arable land, etc may be deplored, the process is being facilitated by transport. Nevertheless it has no option, as it is doing no more than meeting its functional responsibility of catering for the demands of the people as they arise. This work has so far examined in broad principles the mechanics of how that responsibility is being met. The physical and environmental effects of its achievement will now be considered.

LAND SPACE AND POPULATION

Space (in terms of adequate space to move around) and the associated environmental problems of the day are attributable to the presence of people. Some might say there are too many people even though in parts of the developed world the birth rate may be falling as, for example, more women are putting careers before family. Nevertheless, overall, the world's population is increasing. Statistics and predictions are contained in Table 8.1 and

Table 8.1 World population statistics and predictions split between the developed and the developing regions.

		Population (millions) Developed regions 000	Developing regions 000
1950		813	1711
1960		916	2111
1970		1007	2694
1980		1083	3365
1990		1148	4134
2000	H	1192	4931
	M	1187	4904
	L	1183	4879
2010	H	1231	5829
	M	1206	5684
	L	1186	5539
2020	H	1268	7294
	M	1219	6453
	L	1167	6097
2030	H	1297	7802
	M	1212	7159
	L	1121	6504
2040	H	1320	8803
	M	1189	7741
	L	1047	6699
2050	H	1352	9805
	M	1162	8205
	L	959	6703

H = High Variant   M = Medium Variant   L = Low Variant
Note: The two columns together give the world total figure

Source: UN World Population Prospects. Annex 1. Demographic Indicators. 1996 Revision.

the information is depicted in graphical form in Figure 8.1 from which it can be seen that on the higher variant, the population of the developing world is predicted to double between 2000 and 2050! In the developed world figures are static or falling away but it can make little impression in

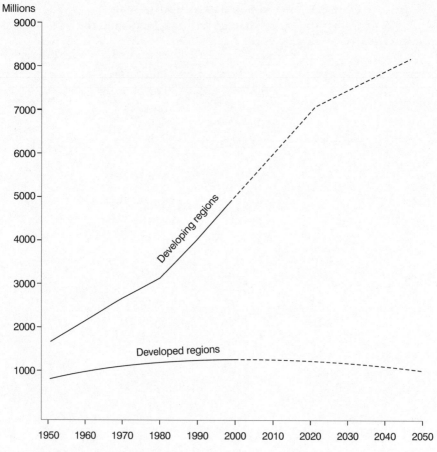

Millions

Figure 8.1 Graphical indication of world population trends.

terms of world totals. This of course begs the question: to what extent can such a population explosion be sustained? The US Population Institute is reported as saying that nearly 98 per cent of the annual increase in population occurs in the less developed regions by those least able to support their growing millions and withstand the consequences of runaway growth, economic stagnation, hunger, malnutrition and urban deterioration.

Regardless of these increases, once people are here they require 'space' in which to live. They need land for their dwellings and for their industry and also room to move around. This is an important issue, distinct from the pollution aspect discussed below. What is important in this context, however, is that one of their needs is adequate communication. But their conveyances occupy roads, roads occupy land and the overall extent of that need is commensurate with their numbers. But the one thing that will not increase is land availability. The surface of the earth is not grow-

Plate 1: A British motorway showing in the immediate foreground four lanes in each direction. To illustrate the extravagant landtake of road transport compared with rail, UIC is on record as saying that just a double-track high-speed railway line would be sufficient to accommodate the equivalent of the maximum traffic that could be carried on a dual-carriageway road with four lanes in each direction, as is pictured here.

Plate 2: Residential roads are used as parking lots with the result that traffic can no longer circulate. If a vehicle stops for any reason, such as to pick up or set down, to deliver or to collect household rubbish, then there is a complete blockage which even the emergency services could not penetrate. This residential road in north-west London (England) illustrates the point.

Plate 3: Cycling is prohibited along the promenade at Southbourne in Dorset (England) yet the arrogance and total disregard of the law by cyclists without any consideration for pedestrians have made this scene a part of everyday life. The local authority continues to spend money on the erection of prohibition signs whilst knowing that there is no intention of enforcement.

Plate 4: At Iford in Dorset (England), as is the practice nationwide, traffic-calming measures have been introduced which have been successful in forcing motorists to drive more slowly. But cyclists prefer to use the footpath even though special provision has been made for them (between the centre bollards and the kerb). Here two cyclists have just come round a blind corner at speed. Pedestrians know better than to walk that way.

Plate 5: In Beijing (China) there is much commuting by bicycle as this cycle park shows. But the local cyclists know better than to use the footpath for this purpose and it is, therefore, pleasant to walk.

Plate 6: In Australia it is national policy to combine cyclists and pedestrians on a single way, as this sign shows. But make no mistake: in Australia it is safer to walk along the road than it is to walk on the path. You can hear the cars coming! Little children are particularly vulnerable. They only think they are safe.

Plate 7. In parts of the Far East road space is not wasted although it is not always used for the purpose that it was originally intended. This is a typical highway (underneath an elevated LRT line) in suburban Manila (Philippines). A close inspection will reveal traffic endeavouring to struggle through.

Plate 8: Although China is a densely populated country, roads are often wide, as they are here in Xian, and traders do not overflow on to them. Some stalls can be seen on what is a wide footpath on the opposite side of the road between the articulated bus and the horse and cart. The cyclist is correctly using the road with the footpath reserved for pedestrians.

Plate 9: The main form of local transport in Ho Chi Minh City (Vietnam) is by low-powered motorbike. Although there are acceptable local bus services, they are not widely used as the motorbikes are superior in terms of convenience and cost if not in comfort. The main thoroughfares are wide with the centre lanes reserved for four-wheeled vehicles (and which are mostly empty). The two side carriageways (or service roads) are used by the motor cyclists (and bicycles). In the peaks they travel five abreast nose to tail at about 20–25mph with up to about 300 passing within the space of one minute. Atmospheric pollution and safety in road crossing both leave something to be desired.

Plate 10: Parking is prohibited along the right-hand side of the High Street in Christchurch (England) except that tolerance was shown to handicapped people who are given special stickers. But as this illustration shows, this special concession just cannot be. The road on this occasion was totally blocked by their presence and the local authority has had to take firmer prohibitive measures to curb this effect.

Plate 11: It is useless providing bus lanes unless they are enforced. Here in an inner suburb of London (England) a motorist has parked in a bus lane with emergency lights flashing whilst he is in the telephone booth and the bus cannot pass. Not only is it illegal for a car to enter the bus lane, parking is prohibited in this area anyway.

Plate 12: Whilst taxis may be convenient, when they proliferate they create congestion and unnecessary pollution. They cause delay when they make U-turns in the road (which they often do) and when they stop to pick up and set down passengers regardless of other traffic, particularly when setting down and there is a discussion over the fare. In busy areas they will have to be controlled. In this view of central London (England) there are as many taxis as there are buses and not one that is facing the camera has a passenger on board.

Plate 13: Large goods vehicles may be economical and this one is about to deliver to a supermarket where there are facilities for off-street unloading. They provide an essential service but, unlike buses which have set routes and stops, they are less than ideal for busy and congested urban streets as they are for narrower roads and villages outside the built-up areas. There is an alternative, in that deliveries could be made at night and this may have to come. Then there will be complaints about noise.

Plate 14: As in Plate 13, large vehicles are economical and, again, delivery is being made to a supermarket. But here in Cairns, Queensland (Australia) the streets are wide and there is no traffic congestion. If the off-street unloading bay is occupied then there is no problem in standing on the road.

Plate 15: There is no need for car restraints here in Rockhampton, Queensland (Australia). Even the railway shares the public highway within the city limits but the basic passenger timetable is only five trains in each direction per week, all of which pass through Rockhampton at night. There are, however, some goods trains which do traverse this section of track during the day. It is the main line from Cairns south to Brisbane. In traffic terms there might be a case for a through service to Sydney but at Brisbane there is a break of gauge. The track in the picture is 1067mm but from Brisbane southwards it is the standard 1435mm gauge.

Plate 16: There is even less need for car restraints on this road to Charters Towers in Queensland (Australia). It was said in the text that restrictions on cars should be selective and here the need is not likely to arise for a long time. For the next 100 miles onwards it is unlikely that another vehicle will be encountered in either direction. Facing due south, the populated east coast lies about 80 miles across the bush but to the west it is about 1600 miles to the sea and there is virtually nothing in between.

Plate 17: In Karachi (Pakistan) midibuses and three-wheel motorized rickshaws (taxis) are in abundance, neither of which do anything for the improvement of atmospheric pollution, traffic congestion or comfort. On the buses, male passengers use either entrance but occupy the compartment to the rear of the forward entrance. Female passengers use the forward entrance only and occupy the smaller space to the front. Westerners seldom (if ever) use these buses which means that they are left with a somewhat perilous ride in a motorized rickshaw (in some parts of the Far East, vehicles of this type are known as 'tuk tuks') with the only other option being a conventional taxi.

Plate 18: The interior of a midibus in London (England). Cramped conditions such as this are just not good enough to encourage people out of their cars. Bus companies will have to do something better.

Plate 19: If this was not a genuine photograph it would be difficult to believe the congestion (and pollution) that minibuses can cause. Here in downtown Manila (Philippines) every vehicle that is facing the camera is a bus, mostly minibuses (known locally as 'jeepneys') carrying about 16 people. Had double-decks been used, the entire traffic between the camera and the bridge in the background in the facing direction could have been accommodated in about six vehicles.
Reproduced by kind permission of the National Book Store, Manila

Plate 20: Until comparatively recently, Manila (Philippines) had some double-deck buses but it was alleged that the roads were not suitable for their size and that there was difficulty in obtaining spare parts. This ultimately led to their demise and now the jeepneys cater for most of the traffic. This is an unfortunate development, particularly in terms of atmospheric pollution. Physically, the roads are suitable but there was a problem of congestion. The way to solve that, however, would be to have more double-decks. But things have not gone that way and the few double-decks that were there have been found what some might regard as 'better uses' as is illustrated. If pollution is ever to be tackled then policies will have to be different. When the elevated LRT line was constructed, a ground clearance allowed for double-deck operation.

Plate 21: Elevated railways are a cheaper construction than are underground lines but they are visually intrusive and overpowering along what is here a busy shopping street in Manila (Philippines).

Plate 22: Some long-distance coach services in Australia also carry goods. Greyhound (Australia) operates a parcels express door to door service and special trailers are used for this purpose.

Plate 23: This shopping street in Melbourne (Australia) has been pedestrianized but the tram remains and sits comfortably in the centre of what was once the road. Had there been no trams it would have been tempting for the planners to put the buses elsewhere but in this case passengers continue to enjoy a public transport facility to the place where they want to go.

Plate 24: One of the largest passenger-carrying vehicles, this bus of Kowloon Motor Bus in Hong Kong (China) has over 100 seats. With standing passengers the total is about 170. It would take four midibuses to give that level of accommodation (with four drivers and four diesel engines) and it is ideal for heavily used services in urban areas where demand is sufficient to justify a frequent timetable. Large vehicles are more economical providing it does not mean carrying around a large number of empty seats during the course of the working day.
Reproduced by kind permission of Duple (Metsec) Ltd

Plate 25: Up until the 1930s in Great Britain and elsewhere tradesmen made their rounds with vans or horses and carts along local residential streets, selling all items of foodstuffs along with a number of other commodities. Housewives made their purchases at their front doors at what were in effect mobile shops and no carrying of heavy items from more distant shopping centres were involved. Those facilities have now virtually disappeared and in any case the traders would have difficulty today in penetrating the masses of cars that are parked along the side roads. However, the result is that the need now arises to convey the contents of large shopping trolleys from supermarkets or wherever to the home and that is why so many women of today need a car, whereas their grandmothers did not. In Great Britain an element of the milk delivery service still survives and the electric vehicle (which replaced the horse and which is pictured here) is ideal for the purpose as the required range and speed is not great and there are frequent stops and starts. There has been much development of batteries since this concept was first introduced but still further work needs to be done before this form of propulsion becomes an acceptable alternative to the car.

Plate 26: The notice at the foot of this bus-stop flag in Putney High Street, being a suburb of London (England) reads 'On Mondays to Fridays buses do not stop here from 8am–9.30am & from 4.30pm–6.30pm'. There is a popular supermarket adjacent to this stop and the afternoon period is a time when women congregate there with heavy shopping bags. It is a long walk to the next stops either way and both directions entail crossing busy roads. On the other hand, taxis are permitted to stand in an adjoining side road and when hired they add to the medley in the High Street. But buses are not allowed to stop as they would delay the cars! The authorities will have to do better than this if there ever is to be a swing to public transport. Note: this restriction has subsequently been lifted.

Plate 27: Although there are isolated examples of monorails which form part of an urban public transport network they are used mostly as local distributors in large shopping and/or recreational complexes. This example in Sydney, New South Wales (Australia) which is of the straddle type provides a circular route around the exhibition, conference and entertainment centre of Darling Harbour and gives a link to some of the major hotels.

Plate 28: Not many conventional street tramways have survived to this day although some have been upgraded either in whole or in part to LRT standards. The example pictured here continues to provide a frequent and heavily used service which runs parallel with the northern shore of Victoria Island in Hong Kong (China). This particular line has a comparatively narrow gauge of 1067 mm.

Plate 29: A rack (or cog) system may be used on railways where the gradients are too steep to permit normal adhesion. This is an example of a short line that runs up to the high altitude suburb of Fogaskereku in Budapest (Hungary). A cog which is fitted on the train (which in this instance is electrically powered with overhead catenary) engages the teeth of the centre rail which can be seen in the foreground.

Plate 30: The provision of separate classes of travel is a contentious issue. It may irritate but in some cases it is essential. On this boat which provides a service along the Yangtze river in China there is no designated 1st class as such but there is a part reserved for a '2nd class'. The general accommodation would be just a bit too basic for some people and if something better was not provided they would not be prepared to travel. Here, the majority may be envious as they peer wistfully through the bars at the 'superior' passengers on the other side.

Plate 31: The trains of the suburban railway system of Sydney (Australia) are double-deck and are operated by the State Railway Authority of New South Wales in consultation with the State Transit Authority which runs the buses and the ferries. It is an electrified system with overhead catenary and in common with the rest of the New South Wales railway network, it has a standard gauge of 1435mm.

Plate 32: The distinction between a tram, light rail transit and a full scale railway (or heavy rapid transit) is always blurred but this is clearly more than a tram. It is the elevated LRT line in Manila (Philippines) which runs on a north–south axis through the city centre. Although Philippine National Railways has a narrower gauge of 1067mm, the LRT line (which is a separate undertaking) has been constructed to the standard gauge of 1435mm.

Plate 33: This is a designated light railway (the Docklands Light Railway) in London (England). Conforming to the standard gauge of 1435mm, it serves the redeveloped areas of what were once occupied by the London Docks; it is a fully automated system using much of the redundant and derelict railway track that once serviced the docks. It is electrified on the third rail principle with the third rail raised above the trackside. At this particular point the line runs alongside a conventional main line railway which is also electrified but with overhead catenary.

Plate 34: This picture illustrates the shallow tunnels of the 'cut and cover' system used for the construction of many underground railways as it was in the first instance here in London (England). Electrification by the third and fourth rail arrangement is used on London Underground (the running rails are utilized also for automatic track circuit signalling purposes) which has a standard track gauge of 1435mm. The sub-surface tunnels are built to accommodate rolling stock with a standard loading gauge. A train is seen approaching Barbican Station.

Plate 35: The first underground railway in the world was built in London (England) in 1863. It ran from Baker Street to Farringdon and this is Farringdon Station where the line can be seen disappearing into a shallow tunnel. Trains on the sub-surface lines were originally steam-hauled but were electrified around 1900. However, the train in the picture is part of the national railway system on a through running arrangement. These longer distance trains, although also electric, work to a different system. On the southern section for historical reasons the tracks are electrified on the third rail principle whilst on the northern section it is with overhead catenary. Trains are therefore dual-powered and here one is on London Transport track but with high voltage overhead catenary. It is the changeover point between the two electrified systems.

Plate 36: Note in particular the wheels of this train and the track (which is electrified on the third rail principle). The actual running tracks are designed for the pneumatic rubber tyres with added guide wheels and rails. The conventional steel rails and hidden steel wheels on the train are used for emergency purposes and guiding through switches. The rubber-tyred train is a French concept developed in the 1950s to combat the disadvantages of the lesser resistance to rolling which is a feature of railed transport (and was why horses could pull heavier loads on trams than on buses). The rubber tyres, as well as giving a very smooth, quiet ride, have more grip which enables better acceleration and hence speed over short distances. These trains can also tackle steeper gradients which facilitates route planning and construction as different levels can be achieved within shorter distances. This train of RATP is pictured in the shallow underground Gare de Lyon station in Paris (France).
Reproduced by kind permission of RATP

Plate 37: By the turn of the nineteenth century, London (England) had moved away from the 'cut and cover' principle of underground railway construction and began deep level burrowing. It is these tunnels that are known colloquially as the 'tube'. Here a 'tube' train of London Underground (London Transport) which closely fits the bored tunnels is seen at Charing Cross Station in London.
Reproduced by kind permission of Jubilee Line Extension Project/QA Photos

Plate 38: Burrowed tunnels are expensive to construct: the bigger the bore the greater the cost. At the time, therefore, a decision was taken in London (England) to opt for a smaller bore and hence the loading gauge of a 'tube' train is smaller than standard although the track gauge is the same at the standard 1435mm. This interior view of a train illustrates the lesser headroom that is available on the rolling stock that is able to run through the deep burrowed tunnels.
Reproduced by kind permission of Jubilee Line Extension Project/QA Photos

Plate 39: Tourism makes its mark. This 'typical' village of the White Karen Hill Tribe in northern Thailand is on the line of an organized itinerary for tourists to enable them to see 'how the natives live'. Such visits have been made possible through easy access – by air from the West to the main centres and then by coach along roads that have been suitably widened to accommodate the larger vehicles. The village has, of course, become stage-managed. The locals have their eyes on a new source of income, the village is becoming rather less rural and rather more artificial and the environment is changing. With constant contact with people from the outside world the locals will soon no longer be content to maintain their natural way of life. But it is the tourists that bring in the money.

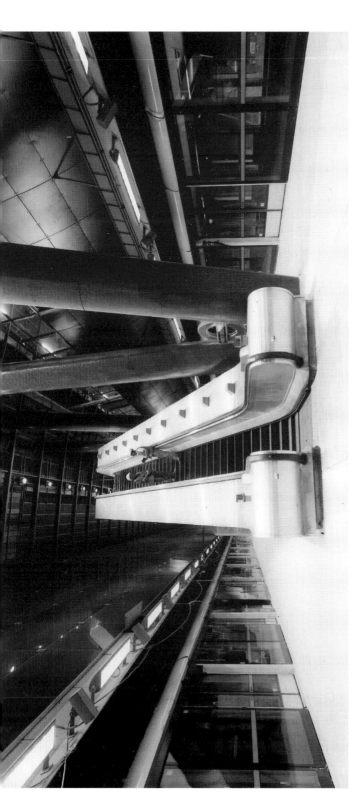

Plate 40: One of the most recently constructed stations of the London Underground network, North Greenwich (opened in 1999) is an open cut station as opposed to mining. The station is built in a vast open box over 13,000 feet (400m) long, 105 feet (32m) wide and 82 feet (25m) deep with banks of escalators leading to the surface. As can be seen, platform edge doors have been provided which minimizes draughts and is an additional safety feature which prevents people jumping, falling or being pushed into the path of a train. When a train stops its doors align with the barrier doors and both open simultaneously. This is a new departure for London but the system has already been working with success in Lille (France) and Singapore. There is also another important transport feature of this station. It is built in one of the few remaining pockets of London which was hitherto completely undeveloped. It lies in a bend of the river on land that is composed of sands and silts too unstable for tunnelling until the arrival of modern technology. But, more importantly in the context of the principles of transport, the new underground line is acting as a catalyst for the development of the area. What was once the site of a gasworks, next door to the new Underground station is also a new bus station, the Millennium Dome is coming and eventually there will be office accommodation, shops, leisure facilities and 5000 homes. Without the railway this would not have happened.

Reproduced by kind permission of Jubilee Line Extension Project/QA Photos

Plate 41: One of the earliest railways in the world, the London and Greenwich Railway which opened in 1836, was elevated as the site, being adjacent to the River Thames, traversed what was at that time marshland and was liable to flooding. The elevated structure extended for 3¾ miles (6km) and involved the construction of 878 brick arches. Even then Parliamentary approval was necessary for construction and the project did involve slum clearance. However, there was no effect on important property owners and the distress that was caused to the local residents was apparently of small concern. As a matter of interest, people subsequently moved in to live underneath these structures in abject poverty and their plight was immortalized by the late Flanagan and Allan with their popular duet 'Underneath the Arches'. However, these spaces are now used for commercial purposes and the more prosperous incumbents park their private cars and vans in front. The railway above is now a part of the national railway network.

Plate 42: The State Railway of Thailand has three classes of travel but 1st class is now largely confined to night sleepers. Although most trains have mixed 2nd and 3rd class accommodation, a few have only one or the other. This 3-car diesel multiple unit 'Sprinter' is 2nd class only and provides the fastest day service from Bangkok to Chiang Mai (Thailand). However, notwithstanding its name, it does not do a lot of sprinting. For the 466 mile (751km) journey it has a scheduled time of 10 hours 40 minutes without any lengthy intermediate stops and it is known to arrive late. It is here seen being serviced *en route* at Phitsanulok. Note the metre gauge track, much of which consists only of a single line. This inhibits punctual operation as any delay to a train in one direction is transmitted through to trains travelling the other way.

Plate 43: In the right conditions, transport by inland waterway is characterized by low cost, high energy efficiency and low adverse environmental impact. It is especially suited for the movement of bulk and semi-bulk goods over comparatively long distances. In England, British Waterways has redeveloped its Sheffield and South Yorkshire Navigation to enable larger vessels to navigate through from the Humber ports to Rotherham which is some 80 miles (129km) inland from the open sea. Here a push-tow train is seen at Conisbrough on the SSYN.
Reproduced by kind permission of British Waterways

Plate 44: In some of the larger urban areas, river boats make an important contribution to the local passenger transport network. Although efforts are again being made to develop such services in London (England), the River Thames is not ideal for this purpose as it meanders. This makes journey times slow and stopping points are not always at the most popular traffic objectives. This vessel is seen working on the service from Westminster to Greenwich but it does not compare with road or rail in terms of time, cost, frequency or convenience and it is used almost entirely by tourists.
Reproduced by kind permission of the Port of London Authority

Plate 45: A general view of docks suitable for ocean-going shipping at Tilbury (England). Tilbury is a large 'forest' terminal and in the foreground timber can be seen on the quayside having recently been unloaded. To the rear is a container ship with containers being stacked on the vessel. The specialized cranes and further containers waiting to be loaded can be seen in the stacking area on the left hand side.
Reproduced by kind permission of the Port of London Authority

Plate 46: Thames Europort at Dartford (England) which deals with container and roll-on-roll-off vessels on the short sea routes across to the Continent. On the left mobile cranes are unloading containers whilst on the right road haulage vehicles can be seen entering the ro-ro vessel through the stern doors.
Reproduced by kind permission of the Port of London Authority

Plate 47: Container terminal at Southampton (England) showing gantry cranes and container stacking area. A feature of a container port is the large flat stacking area which in this example is 104 acres (42ha). This port is equipped with nine ship-to-shore gantry cranes which are capable of handling the largest container ships. The terminal can work three panamax-size vessels simultaneously and has the capacity to handle more than 850,000 teus annually. A 'panamax' is the maximum sized vessel that can pass through the Panama Canal which links the Atlantic and Pacific Oceans and a teu is a container measurement, being a twenty-foot equivalent unit (eg 1×20ft = 1 teu and most containers are either 20 foot or 40 foot long). The port is served by both road and rail with regular dedicated train services to all parts of the country and has accreditation by Lloyd's Register Quality Assurance (see Chapter 7).
Reproduced by kind permission of P&O

Plate 48: The bulk carrier *Vine*, dwt (deadweight tons) 123,738. There are several ways of calculating tonnage, each producing very different figures and even these are not universal between nations. The Tonnage Convention of 1994 as adopted at a conference held under the auspices of IMO (see Chapter 7) provides for gross and net tonnages, but the system is not yet universal and the previous classifications are still used. Nevertheless, to the operator the figure is important as it is the basis of assessing harbour and canal dues, etc. Deadweight tonnage indicates the number of tons that a ship can carry which includes not only cargo but also stores and bunker oil. In other words, it is the difference between the number of tons of water displaced when light and when fully laden (ie, when submerged to the loadline). The *Vine* pictured here is used for the worldwide carriage of dry cargoes such as grain for which it has a capacity of 136,041.5 cubic metres and requires a crew of 25. It has a speed of 14 knots.
Reproduced by kind permission of P&O

Plate 49: A roll-on roll-off passenger ferry boat *Stena Empereur*, 28,559 grt, working on the Dover (England) to Calais (France) service. It has a speed of 19 knots with a capacity for 2036 passengers, 550 cars and 78 trailers and requires a crew of 86. For tonnage measurement purposes the passenger ships are here expressed in gross tons, being a figure related to the ship, not the cargo. In simplified terms the gross registered tonnage is calculated by dividing by 100 the volume in terms of cubic feet of the vessel's closed-in spaces.
Reproduced by kind permission of P&O

Plate 50: The P&O passenger ship *Grand Princess*, 108,900 grt. Among the largest of the world's passenger ships (the length and breadth dimensions being slightly larger than the bulk carrier pictured at Plate 48), it has a speed of 22.5 knots and carries 2600 passengers. Built for the cruising market (which is a growth industry) and with a particular eye on North America, the vessel encompasses multiple dining and entertainment locations, 24-hour restaurants, virtual reality arcades, video studios and wedding chapels. In conformity with current trends, it has maximized on private balconies. A crew of 1060 is indicative of the service offered – a figure which is interesting to compare with the vessels depicted at Plates 48 and 49.
Reproduced by kind permission of P&O

Plate 51: A catamaran type twin-hull vessel of Stena Line designated 'HSS' which stands for Highspeed Sea Services. Vessels built on this principle have less drag and hence are faster than conventional boats. Powered by waterjets via four gas turbines, its average speed is around 40 knots compared with less than 20 knots of the standard ferryboats. The voyage from Holyhead (Wales) to Dun Laoghaire (Ireland), for example, a distance of some 60 miles (96km), is scheduled for 1 hour 39 minutes. Its special design has reduced the pitch and roll of conventional ferries and of catamarans. The HSS craft pictured here has a capacity for 1500 passengers and 375 cars (or 50 trucks/trailers plus 100 cars). With vessels at sea, size is deceptive. To put it in perspective, the overall dimensions of this HSS craft approximate to that of a football pitch.
Reproduced by kind permission of Stena Line

Plate 52: A hovercraft of Hoverspeed working on the Dover (England)–Calais (France) service. The hull of a conventional boat floats in water and it is necessary for the engine to overcome drag. With hovercraft the unit is lifted clear of the water which substantially reduces the drag and thereby permits higher speeds. The journey from Dover to Calais, a distance of some 21 miles, is accomplished in 35 minutes. There is accommodation for 55 cars and 424 passengers. The hovercraft functions by generating a cushion of air under pressure from fans that raise the craft a few feet and thereby enable pusher propellers to thrust it across the water (or ground) surface. Its flexible skirt allows greater hover heights and as the craft lifts the skirt hangs beneath it to retain a cushion of air deep enough to negotiate waves on the sea or obstacles on land. *Reproduced by kind permission of Hoverspeed*

Plate 53: A hydrofoil in Hong Kong harbour (China). Like the hovercraft, the unit is lifted clear of the water which reduces drag and thereby permits greater speeds. But there the similarity ends. In the case of the hydrofoil it is the action of the foils (or 'wings') that creates the lift (it 'flies' in the water) and when approaching a landing stage the foils are released and the craft then functions as a normal boat. In this instance, the foil at the front which is lifting the hull can be seen. Unlike the hovercraft, the hydrofoil is not amphibious.

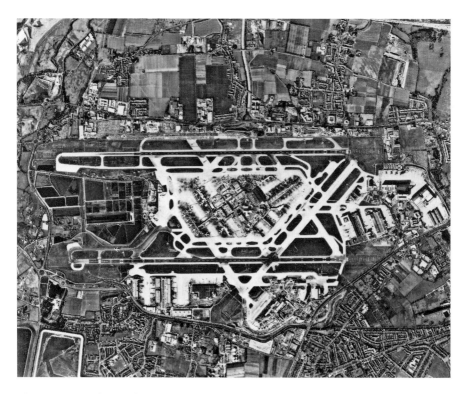

Plate 54: An aerial view of Heathrow Airport (London, England). The terminal complex which contains Terminals 1, 2 and 3 lies in the centre of the picture surrounded by two parallel east to west runways north and south of the centre with a third runway located diagonally on the eastern side. Terminal 4 lies outside the 'inverted triangle' at the south-eastern extremity. The main approach roads are clearly visible. The link from the national motorway network feeds in from the north (in the top centre of the picture) and then tunnels under the northernmost runway into the central terminal area. Within this centre area there is a bus and coach station and two railway stations, one of London Underground (rapid transit) and the other of a fast non-stop service into London of the national railway system. Situated some 15 miles from central London, the airport area covers nearly 3000 acres (1200 hectares) and the longest runway is 2½ miles (4km). In the year 1997/98, more than 58 million passengers were handled at Heathrow with up to more than 1000 aircraft movements in a day. The result is that the facilities are over-stretched and a fifth terminal is required. If planning permission is granted, Terminal 5 would be located in the darker area in the left centre of the picture which is at present used as a sludge works. The locations of the various features can be more readily identified by reference to the skeleton drawing on the next page.
Reproduced by kind permission of BAA

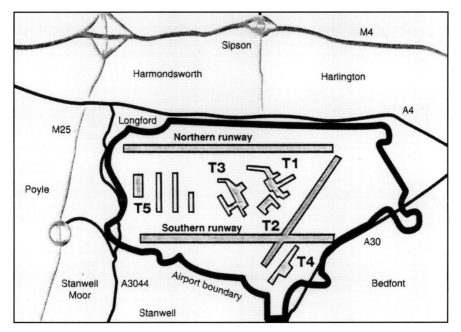

Diagram of London (Heathrow) Airport (see Plate 54).
Reproduced by kind permission of BAA.

Plate 55: A Boeing 747–400 aircraft of KLM Royal Dutch Airlines. This is an enlarged version of the original B747 (known colloquially as a 'jumbo jet') and can be recognized by its stretched upper deck and the angled winglets at the tip of each wing. The winglets reduce the drag due to airflow which, combined with greater fuel efficiency, gives the aircraft a wider range and greater payload. The actual passenger capacity varies according to the seat configuration and allocation of the different classes but the aircraft illustrated has accommodation for 424 passengers. It has a maximum range of some 8000 miles (12,600km) and an average cruising speed of 575mph (920kph) at an altitude of about 45,100 feet (or about 8 miles or 13,747 metres). Whilst a few aircraft (including B747s) are adapted entirely for the conveyance of air cargo, the passenger craft do also carry freight and mail and the maximum overall take-off weight of the B747 is approaching 400 tons (390,100ks). Operationally, there must be a proper control over weight in relation to passengers and freight and with the fulcrum point and centre of gravity at a predetermined place it must be distributed correctly to ensure balance and stability in the air. There is always a need to achieve the maximum payload, be it passengers or cargo, and to the operator it becomes a question of mathematics, allowing generally 170lb (77kg) per passenger, whatever the potential cargo might be and the fuel. The actual amount of fuel required for a flight has to be determined bearing in mind that delays whilst taxing consume extra fuel which might not be burnt off before take-off and for which there is a maximum safety weight anyway. Diversions, such as for weather, take more fuel and planes should not land with an excessive weight as this necessitates a faster landing speed which means longer running on the ground (with an adequate length of runway) and more braking power. The fuel, therefore, involves a series of considerations for which B747s have a capacity of over 57,000 gallons (216,000 litres).

Reproduced by kind permission of KLM Royal Dutch Airlines

ing and neither can it grow. On the contrary, if the ice caps of the polar regions melt, as is predicted, as a result of global warming which in turn is a result of people living up to their now accustomed quality of life, land availability will diminish as the seas rise and begin to engulf our shores. Thus a static land availability at best or a shrinking land mass at worst cannot forever sustain a constantly increasing population. Returning to the transport syndrome, the corollary here is that there is a physical limit to the number of roads that can be built (and aeroplanes that can fly) and the amount of traffic that they can carry.

To date, population overspills and the quest for a better way of life are being met (apart from a very small amount of land reclamation as in Holland) by emigration and the environmentally damaging encroachment on to arable land, wildlife habitats and forested areas. An example is in Kenya where the white rhinoceros and the pink flamingo are in danger of extinction. The reason is that their habitats are being occupied by people who must find places to live and transport is one of the factors that enables them to do that. But the white rhinoceros and pink flamingos are popular tourist attractions. People come to see them. Without them they will not come. The inhabitants need land as a place to live but for economic reasons, Kenya (like many other places (see Chapter 2)) needs and relies on tourists. Where do you go from there? At some stage there must be a balance between these conflicting demands.

All of this is serious enough. But as the earth's resources (which includes land space, parts of which are inhospitable anyway) are finite, it is clear that something, somewhere, sometime, will have to give. In other words, whatever the circumstances, it is inevitable that there is a physical limit to the number of people that the world can support.

It was noted in Chapter 1 that the availability of transport influences much of what happens. It has set patterns and it dominates our daily lives in countless ways. But human reproduction is not one of its attributes. Too many people in too little space is not a transport issue and experts of a different faculty must face that dilemma. China, for example, has introduced penalties in some areas for families who have more than one child, on the grounds that a soaring birth rate drains national resources whilst causing famine, poverty and unrest. But where there is no birth control (for any reason) and the population proliferates, there is always the danger that nature will step in and do it for them. Some will say that there is a supernatural power which controls these things; certainly the world has witnessed over recent times a massive loss of human life through hostility, famine, disease and other acts of God. Laurie Garrett, for example, who has undertaken research in this field[1] has found a crop of fatal diseases in addition to Aids, such as Ebola, rodent-born hantaviruses, toxic shock syndrome, Legionnaire's disease, drug-resistant tuberculosis, Lyme disease and flesh-eating bacteria, all being brought about by the need to cater for too many people and hence unbalancing the Earth's microecology. Then, of course, there is always the possibility of biological warfare which could wipe out millions with very little effort in very little time.

The problem – indeed, the impossibility – of too many people in too small a space is illustrated by a study of the future passenger transport requirements of an urban area in one of the developing countries of Africa which was recently undertaken by the author. The future demand was based on already prepared population predictions, as it is appropriate to do. But the estimated vehicle requirements over periods of 10 and 15 years hence were of such magnitude that in no way could they have been physically accommodated either in the available infrastructure or even space on the ground. And the cost would have been phenomenal. In fact, neither could any of the other supporting services be made adequate and that was assuming that places could be found for everybody to live (which was not a part of the exercise). The only practicable approach would have been to create an entirely new town 'somewhere in the bush' which was not intended. However, there was much illness among the people. Aids was endemic. Disease was rampant which would inevitably decimate their numbers. It was recognized that, under the circumstances of the day, it was unlikely that the hitherto accepted population predictions would ever be realized. Then, with a substantial percentage reduction in future anticipated population growth, and hence a lower level of assumed demand, it became possible to prepare a realistic plan.

Worldwide, disasters of this kind are taking their toll alongside medical research which is also an ongoing process. Research and development is being undertaken in the interests of humanity to alleviate misery, prevent disease and prolong life. All very laudable but it begs the question whether medical science can really continue to fly in the face of nature and win in the end. If the lives of future generations are not only to be safeguarded but extended, then another problem will arise. The question then will be how to accommodate the people who would hitherto have died but instead continue to live? The population levels would become swollen to the extent that they could no longer all be sustained within the world's available resources. Furthermore, it would be unlikely that the very elderly amongst them would be making any contribution to national economies anyway. Again that is not a transport conundrum but transport would be involved.

There is a sequel and what is a more relevant scenario in this context: congestion on roads. In one sense, if any credence at all is to be given to these thoughts, the problem of space will never arise as nature will take care of the longer term. But nature may not be thinking specifically about today's urban traffic congestion. Regardless, therefore, of what has just been said, this short-term aspect must remain in urgent need of attention and if there are any more breakthroughs in combating disease, then that problem will only be exacerbated. The matter is discussed further under the sub-heading 'The private car' below.

LANDTAKE AND LANDSCAPE

The environment has become a popular topic and it is fashionable to pay at least lip service to its preservation at the expense of transport infrastructure and facilities (and also to virtually everything else). Although the benefits of transport facilities are readily accepted, any resultant, albeit inevitable, environmental unfriendliness is not. The need for adequate transport cannot be challenged but its provision involves infrastructure which must occupy land and is liable to scar the landscape. As was said in Chapter 3, roads, railways, seaports and airports are all extravagant to varying extents in this respect and are frequently the subject of complaint. But individuals have their own pet fancies and priorities according to their personal preferences and ways of life. There is no common consensus on what constitutes 'a good environment' and the following case studies illustrate the point.

The first example cites Cairns in Queensland, Australia. Cairns was a seaside village adjacent to the Trinity Inlet when, around 1912, dredging to make a deep-water channel enabled ocean-going shipping to dock at the local wharves. Along the waterfront away from the inlet there was a sandy beach. Subsequent to dredging, the sand gradually gave way to mud and in due course the area became inhospitable and useless for recreational purposes. This was an example where a ruined environment was placed firmly at the door of transport, or so it was said at the time, and the resultant public outcry persists to this day. More recently and to appease the local populace, the matter has been investigated. However, the findings were that the transformation was not necessarily the result of dredging; it could have been caused by some natural phenomenon. The area is flat and cyclones or the loss of mangrove swamps which do encourage sand around them could equally have been responsible. The case against the port was not therefore proved. But to return to the point of the story: the present generation is still clamouring for the return of its beach. 'Let us enjoy the sand and the sea where we can relax and our children can play' is the plea. 'We need to improve our quality of life – that is what the environment is all about.' Accordingly, suggestions were promulgated to build a new walkway by the sea, remove the mud and replace it with sand. But, 'no no no' says another brand of environmentalist. 'There are pelicans, terns, spoonbills, sandpipers and a host of other birds that have made this their habitat – if you do this the people will come and the birds will fly away – it is their life and their mud and the environment must be preserved'! The point of this narrative is twofold: not only is transport being blamed, rightly or wrongly, for anti-environmental activity; but also what is considered to constitute an acceptable environment is open to different interpretations.

A second example is in Hong Kong, China, where a new airport has been constructed 16 miles (25km) to the west. It lies next to Lantau Island on the site of Chek Lap Kok Island which has been erased from the face of the map to make way for the airport. Its construction has made such a profound impression on the environment and the general scene that it would be difficult to find another transport development that has had a greater impact. Land has been reclaimed from the sea; a new railway link has been provided and a new bridge constructed. Mammoth feats of civil engineering have been effected. A virtual new town has appeared on what was originally an almost uninhabited island seascape which has become well and truly scarred. But the rewards are great. An adequate point of entry and exit to and from southern China is of tremendous strategic value and Kai Tak, the airport replaced, had become grossly overstretched. Furthermore, being adjacent to high-rise blocks and with the runway protruding into the sea, for take-off and landing purposes the original site left much to be desired. Environmental change there certainly has been and the new landscape will not be as tranquil or as picturesque as it used to be. But in this instance there surely can be a strong case to support an argument that the end justifies the means. It appears that the main sufferers are some white dolphins whose habitat has been disturbed and who are not appreciating their alternative provision. This example illustrates the point that the environment cannot be the only consideration, however important it might be.

Now it is reported that BirdLife International has expressed concern that premier wildlife sites will be destroyed by transport (and other developments) which are being considered for funding by the European Union. The proposed works include, among others, port development near Le Havre in France, a road bridge across the Tagus Estuary in Portugal and a mountain railway in the Cairngorms in Scotland. Good for transport – but not good for the birds.

THE PRIVATE CAR

Much of the way of life today revolves around the private car. Some people would say that they could not live without it regardless of the fact that their forefathers had no such private mobility and relied on public transport. Nevertheless, the car is now not only an idol, it is also a menace and for this reason it is appropriate here to make a specific examination of this form of transport.

Environmentalists and pressure groups are castigating the private car (other than when they are driving one). They point to both traffic congestion and atmospheric pollution and advocate a substantial switch to public transport, bicycles and walking. They are of course correct. Congestion on the roads is now causing gross inconvenience and misery and road transport is responsible for much of the pollution. In suburban areas of larger towns, residential roads are used as car parks, day and night, and are sometimes impassable even for the emergency services. In Great Britain, designated national parks, being areas of outstanding natural beauty, have become national car parks. There is a body of opinion that depicts the car as a polluting monster and that it bears much of the responsibility for global warming. Whilst these sentiments can be understood, maybe there is another side to it. The allegation justifies further examination.

Interestingly, in a search for measures to reduce greenhouse gases in the transport sector, Dr Leo Dobes has produced a treatise on this subject on behalf of the Bureau of Transport and Communications Economics based in Canberra (Australia).[2] It is illuminating and contains considerable detail on greenhouse gas emissions in Australia in 1900. With Dr Dobes' thoughts in mind, the writer has applied some figures to London (England).

In the horse era and by about 1900 there were around 4000 buses and trams in London. At two horses per bus or tram and over a 14-hour vehicle day, some 50,000 horses were required and these horses together produced a daily deposit of about 1000 tons of dung. But that was just for public transport. There were (by horse and cart) the collection and delivery services of the railways, road hauliers, the Post Office, local tradesmen, etc which might have added, say, another 20,000 horses. Imagine now the situation as it might have been if mechanical propulsion on the roads had never materialized. It has been said that whilst transport has many attributes, human reproduction is not one of them. It is therefore reasonable to suppose that the people that are here today would have been here in any case even if the horse had remained the mainstay of local transport. Just their distribution and places of domicile might have been different. But it is also reasonable to suppose that the demand for local transport would have risen commensurate with population increases and with it, in the absence of the motor vehicle, the daily deposits of dung.

The horse requirements just described appertained to the situation in London in 1900. Consider now the situation as it could have been under these circumstances some 100 years on around 2000. There would by now have been many more than 70,000 horses required within what has since become Greater London. Never mind the butter mountains of the EU; there would have been dung mountains spread over every conurbation across the world and the lakes would not have been wine either! As is noted below, dung produces methane which is a greenhouse gas. The amounts might vary according to the type of feed and how the dung is eventually stored and disposed of but it would clearly have been there in quantity.

Unlike the work of Dr Dobes, in no way is this a scientific exercise and the issue will not be laboured further. Enough has been said to convey some sort of order of magnitude impression sufficient to accept the conclusion of Dr Dobes in respect of his different exercise that to the extent that transport is responsible, global warming would have come anyway and without the help of the car.

TOURISM

It was said in Chapter 2 that tourism boosts local economies. It is a part of modern-day living and brings pleasure to thousands of people. But it is environmentally damaging. Consider the global repercussions.

Look at, for example, the once tranquil picturesque Spanish villages along the Costas which are now filled with hastily erected cheek by jowl square, characterless, concrete, high-rise hotels together with shops which blossom with 'genuine English fish and chips', 'real German wurst', and gaudy souvenirs. Everything is there to attract the tourist seeking sun and sea and with money to spend.

Look at the Hawaiian Islands in the middle of the Pacific Ocean some 1,100 miles (1,800km) from the nearest point on the American mainland. Again, the sun and the sea makes Honolulu a major attraction now that it is easily accessible. American and Japanese tourists alike are there in their thousands. The waterfront at Waikiki is, like Spain, endowed with high-rise hotels. A cruise to Pearl Harbor to see the watery grave of naval personnel whose ships perished at the hands of the Japanese in 1941 is a popular excursion. The trippers spend their money and go, complete with candy floss, popcorn and Coca-Cola, all to the accompaniment of cameras at the ready and raucous music! Without easy transport, nothing of that would be there.

Look at Alice Springs in the Northern Territory of Australia in the outback and at one time a small trading post in the middle of nowhere. But Alice Springs is within striking distance of Ayers Rock, the world's largest natural monolith, and there lies the attraction. Easy access has now brought tourists from Britain, northern Europe and Japan, again in their thousands, to see this unusual and spectacular piece of nature. So much so that the local indigenous population is begging visitors not to scramble all over it. And Alice Springs has risen to the occasion. Modern facilities have materialized and the people are spending their money in the local supermarkets.

The examples could go on. But mass tourism which transport has made possible is changing the face of the earth. Nations prosper from tourism but some could eventually be destroyed by it as in the longer term it becomes self-defeating. With so many visitors, the natural splendours become ruined and the area is then no longer attractive. It begs the question which should be given first consideration – infrastructure and the economics of today or the environment.

ATMOSPHERIC POLLUTION

Introduction

That part of the pollution of the atmosphere for which transport is responsible is now for discussion. However, to make what follows intelligible, the more important of the chemical compounds that play a role in the pollution process and which are relevant to this study, together with their effects, are briefly summarized as follows:

Carbon monoxide (CO)

Carbon monoxide is poisonous to living organisms and in high concentration will kill.

Carbon dioxide (CO_2)

Carbon dioxide plays a dominant role in global warming (the 'greenhouse effect') and any resultant climatic changes.

Nitrogen oxides (NO_x)

Nitrogen oxides is a collective term embracing the various oxides of nitrogen which, for this purpose, need not be dealt with separately. Also associated are hydrocarbons (airborne particles). A secondary pollutant, nitrogen dioxide (NO_2) is associated with tropospheric ozone (see below) and when concentrated can lead to respiratory problems and inflammation of the bronchial tubes. Another secondary pollutant, nitrous oxide (N_2O) is a greenhouse gas. Nitrogen oxides also damage plants and contribute to acid rain.

Sulphur dioxide (SO_2)

Sulphur dioxide is a pollutant, the main source of which is the burning of fossil fuels (such as coal) which have a sulphur content. It can produce respiratory problems and is the main cause of acid rain.

Methane (CH_4)

Methane is a greenhouse gas but only a small amount is attributable to transport. It is also a fuel.

Volatile organic compounds (VOCs)

VOCs are smog-forming gases which emanate from numerous sources not necessarily associated with transport. Those with a transport relationship include hydrocarbons formed during both the combustion of fuel and the evaporation of unburnt fuel and during manufacturing processes (in this context, manufacturing vehicles, etc). They can cause respiratory diseases and increase the risk of cancer.

Particulates

One of the most harmful of the transport-related pollutants, particulates, like VOCs, emanate from a variety of sources. Their presence is concentrated in black smoke through incomplete combustion from vehicle emissions, the major culprits being diesel-powered vehicles and particularly those that are old and poorly maintained. Particulate matter can also be emitted from industry and from power stations. However, although the vehicles which emit black smoke are the most unpleasant they are not necessarily more damaging. Healthwise, it is the smaller particulates that are the most detrimental (not the black smoke) as they can be inhaled and, if small enough, can penetrate deep into the lungs. In this respect, diesel-powered vehicles are suspect.

Lead

Lead is a cumulative poison affecting in particular the brain and nervous system in children.

Chlorofluorocarbons (CFCs)

Chlorofluorocarbons are one of the related parts of synthetic organic compounds which act as a greenhouse gas. As far as transport is concerned, CFCs have been used mainly in vehicles, etc equipped with refrigeration and/or air conditioning units but their use is being phased out.

ANTI-ENVIRONMENTAL ISSUES AND EFFECTS

Introduction

The term 'pollution' is all-embracing. Air, land and water are being polluted by poisonous chemicals through different activities, some of which are the result of the provision of transport services. It has been said that pollution is 'international', meaning that although it is produced nationally, its effects are felt worldwide. As no one nation can control the actions of another, all must suffer and hence the need for some form of international control. This section will consider the detrimental effects of the various pollutants that have been mentioned. It is axiomatic that when it comes to environmental issues, people tend to accept their accustomed way of life without thought for the damaging communal consequences. If questioned about the need for fresh air, uncontaminated food, preservation of the countryside and not too much extraneous noise, the answers, certainly in the developed world, will be positive. But the majority will not welcome any personal inconvenience necessary to achieve those ideals such as (in this context) restrictions on private motoring, even though there is no wish to experience the dire results that are anticipated if no action is taken. This, of course, is assuming always that they listen to and accept the warnings.

Perhaps there is some excuse if people are sceptical about the predicted gloom as, along with the warnings, seeds of doubt are also being sown. Articles have appeared in the Australian press, for example, expressing misgivings about the predicted global warming (to be discussed below) and asserting that the forbidding conclusions have not found universal acceptance amongst scientists. To develop further this thought, although the prediction of devastating weather changes in the longer term has been accepted as realistic by the United Nations' International Panel on Climatic Change (IPCC), this body has also acknowledged that its acceptance was the result of a majority decision amongst scientists and has not been proved in absolute terms. By implication, therefore, there are other scientists who might think differently. One Australian article has even alleged that the whole thing is tainted by politics and is a ploy to obtain money; but again, presumably, without proof. Nevertheless, even if such allegations are without foundation, it is easy to cast doubt into people's minds if the predictions are unpalatable. On the other hand, another school of thought is that although global warming is here and here to stay, it is more likely to be due to natural changes in the climatic cycle than through pollution. Temperatures may therefore be set to rise even faster than predicted, but for reasons not connected with transport.

However, regardless of the global warming predictions, what nobody can doubt is the poisonous fume-laden atmosphere which results from vehicle emissions. And the uncleanliness of some rivers and beaches and parts of the sea is no myth. Neither is the choking smog (see below) that sometimes hangs over Athens, Bangkok, Los Angeles, Mexico City – one could go on – in fact many big cities when the weather is right and often when it is not. It is there for all to experience and it is reasonable to assume that in the absence of any preventative measures, conditions will get worse. Furthermore (and reverting back to a previous section) there can be no uncertainty about the congestion on our roads. The time is rapidly approaching when this issue will have to be addressed with or without global warming. The work must go on.

When studying pollution certain terms used by the meteorologist and the environmentalist require elucidation. The descriptive titles which are more commonly used will, therefore, now be explained.

Ozone

The word 'ozone' is bandied about in both laudatory and deprecatory terms. This gives rise to confusion and some explanation is necessary. Those with a long memory or a historical knowledge of railway advertising in Great Britain might be aware of a poster once produced in the 1930s by a now defunct railway company: Londoners were invited to take a day trip by train to the south coast. 'Out of the Smokezone and into the Ozone' was the message and certainly it was pleasant to get away from the smoke of the City and enjoy some fresh air and sunshine beside the sea (or so the posters had us believe). A dictionary definition of ozone, even into the 1980s, was 'a form of oxygen having a pungent refreshing smell' or, 'invigorating air at the seaside, etc with an exhilarating influence', which supported its alleged delights. Interestingly, the dictionary definition has now been changed to 'an unstable toxic gas with oxidizing properties' but it is still easy to get the wrong impression.

Ozone is something which is formed from fumes in sunlight. It is a secondary pollutant, the formation of which is enhanced in air already polluted with nitrogen oxides and hydrocarbons. It produces respiratory difficulties in that it irritates the mucous membranes and reduces the function of the lung. This tropospheric ozone (ie, in the lower atmosphere) is a contributory factor to the enhanced greenhouse effect which is discussed below. Nevertheless, in one of its forms, although still not a fresh, bracing air, ozone is not only beneficial; it is essential for life on this planet.

Ozone exists in both the troposphere (the lower atmosphere) and in the stratosphere (the upper atmosphere). The lower layer has just been discussed but there is also a thin layer of ozone gas in the upper atmosphere which absorbs much of the sun's ultra-violet light. As it is the ultra-violet rays which are damaging to crops and can cause skin cancer, this protective 'blanket' has a crucial role in the maintenance of life on earth. Its presence has become a topical issue in environmental terms because certain activities are releasing chlorofluorocarbons (CFCs) which appear to be able to penetrate and thereby make holes in this upper ozone layer. This then allows more ultra-violet rays to pass through to the surface of the earth. Put simply, the present problems would seem to arise from there being not enough ozone high up in the stratosphere and too much of it lower down. Unfortunately, the process of adjustment is not equally simple.

Global warming

References have already been made to global warming and the greenhouse effect. A 'greenhouse gas' absorbs or traps heat as do the glass walls of a greenhouse which allow rays from the sun to pass through and then trap some of the heat inside. But globally there is nothing new about the greenhouse effect. It has been around for a very long time. There is a band of gases in the troposphere (which includes carbon dioxide (CO_2), methane (CH_4) and nitrous oxide (N_2O) as well as water vapour which, by acting in the manner of glass in a greenhouse, traps in some of the warmth from the sun. Without it, temperatures on Earth would fall to the extent that life would no longer be possible. Over the years there has always been a balance of gases appropriate for the maintenance of sustainable temperatures as what was being produced by daily life was at the same time being balanced by natural elimination.

It has been said that there are differences of opinion between scientists on the existence and the causes of global warming. However, if the majority view that there is global warming and it is being caused by pollution is accepted, then it is reasonable to say that it is the effects of modern living which is upsetting the balance just referred to. What follows, therefore, will be based on this assumption. The theory is that an enhanced greenhouse effect is being caused by the emission of more of the original greenhouse gases plus tropospheric ozone which was referred to in the previous sub-section. The increased 'greenhouse' layer makes it a stronger retainer of heat which is the reason for global warming. If the resultant predictions are correct, the consequences are that in due course the sea will engulf some of the earth's land area due to the melting of ice in the polar regions, whilst much of the interiors of the land masses away from the sea will turn to desert. It is being suggested that the process has already begun.

Smoke/fog

A combination of smoke and fog, 'smog' has a complex mixture. The main ingredients of wintertime smog are oxides of nitrogen, sulphur dioxide, carbon monoxide and smoke particles which come mostly from vehicle exhausts and the general burning of gas, oil and coal. In cold weather these pollutants can become trapped at ground level by a layer of colder air above which, if there is no wind blowing, makes it difficult for them to escape. In warmer weather particles from the exhausts of vehicles combine with ozone as temperatures rise and produce the photo-chemical killer smog for which Los Angeles (USA) is particularly famous.

Deforestation

Whilst the term 'deforestation' is self-explanatory, the effects may not be quite as apparent.

Trees, and particularly the world's major forested areas, play a crucial role in the maintenance of life. They absorb carbon dioxide and replenish the atmosphere with oxygen. If the forests are destroyed then fewer trees will be available to undertake this essential biological process. But trees also provide a convenient and useful fuel and building material, being wood, which is an incentive to destroy them. If the trees are cut and not replaced and the wood is burnt (and a large number of people, particularly in the developing countries, rely on wood for their fuel) then, not only can they no longer replenish the atmosphere with oxygen, yet more carbon dioxide is produced by the burning process. Nevertheless, the clearance of forests is attractive in the short term even if in global terms it is irresponsible. It is particularly attractive in countries where expanding populations cannot find easily any other land on which to live.

Acid rain

The reader will be aware that the act of breathing involves a process of inhaling fresh air (which includes oxygen) and then exhaling carbon dioxide. As has just been said, the carbon dioxide (which is an acid gas) is in turn absorbed by plants and trees which, as part of life's cycle, replenish the air with oxygen and thereby maintain a healthy balance. As fossil fuels (which includes oil and coal) come from vegetation which became buried thousands of years ago they are, in effect, the product of stored-up carbon dioxide. When these fuels are burnt the (acidic) carbon dioxide is then once again released into the atmosphere.

The term 'acid rain' means exactly what it implies. When acidic gases (which include sulphur dioxide and nitrogen dioxide as well as carbon dioxide) are released through burning fossil fuels in the course of industrial, domestic or transport activities, they rise and then, in addition to the global warming process, mix with water vapour (being the clouds). The resultant rain is then tainted with acid which in turn finds its way into the soil, rivers and lakes. This can be serious as it damages plants and trees and destroys aquatic life which thereby affects other living things including water birds. It also defaces buildings.

ATTRIBUTION AND CONTROL

Although industry generally is responsible for much of the current environmental damage, transport also plays a major part. According to the 1993 Bulletin of the European Commission,[3] transport was then responsible for some 25 per cent of the total EU CO_2 (greenhouse gas) emissions; a figure which could since have risen to nearer 30 per cent. It is now for consideration to what particular branches of transport can the problems be attributed.

The size of the conveyance has also become an environmental issue, albeit in some cases more perceived than real. A particular cause of public hostility is the 'juggernaut', being the large road haulage vehicle. It is unpopular and the objection to these vehicles proceeding at speed along narrow roads and through small villages is understandable. Neither do people living in the vicinity of airports welcome the larger aircraft with their greater noise and vibration potentials. On the sea, the big oil tankers become the subject of public outcry in the event of an oilspill. Nevertheless, regardless of size, in environmental terms it is still the road vehicles that appear to bear the brunt of public concern and complaint. Ocean-going passenger ships are generally looked upon with admiration on the rare occasions that they are seen; aircraft are mainly disregarded except by those who live near an airport or under a flight-path, and the railway is usually accepted other than by those whose property lies alongside the track.

It is the road vehicles that are always in the public eye. They penetrate the neighbourhoods where people live and walk and shop and they are a part of everyday life. The individual noise and emissions of the vehicles concerned might not be worse than that of the other modes but they are there in large numbers and everybody suffers. Nevertheless, there are different facets to the environment which tend to be mutually exclusive. The point has been made about large road haulage vehicles being environmentally unfriendly but in terms of atmospheric pollution and traffic congestion (and cost), their use is beneficial when compared with a greater number of smaller vehicles that would be necessary to convey the same amount of traffic. It was for this reason (and as was said in Chapter 1) that the European Commission has expressed concern that competitive pressures will lead airlines to opt for higher frequencies with smaller aircraft. Strangely, however, the Commission is making no move to curb competition and it also seems content to see the proliferation of small buses where the principle is just the same.

The 1993 Bulletin of the European Commission[4] also contains data on the environmental damage caused by transport within the EC. It notes that atmospheric pollution has increased substantially over the past 20 years and that the emission of carbon dioxide (CO_2), the main 'greenhouse gas', by motor vehicles has increased by 76 per cent between 1971 and 1989, being an average annual increase of 3.2 per cent.

The following table (as extracted from the Bulletin) shows the extent to which the different modes were responsible in 1990:

Mode	Share of emission of CO_2 in per cent	
	%	%
Private cars	55.4	
Road haulage	22.7	
Buses and coaches	1.6	
Total road		79.7
Total air		10.9
Railway passenger	2.8	
Railway goods	1.1	
Total railway		3.9
Total inland waterway		0.7
Other transport		4.8

Other pollutants with their increases over the period 1971 to 1989 include:

- nitrogen oxide (NO_x), which contributes indirectly to the 'greenhouse effect' and hence global warming and directly to acid rain and the build-up of tropospheric ozone which increased by 68 per cent;
- particulates detrimental to health which increased by 106 per cent;
- hydrocarbon (HC) emissions which contribute to the 'greenhouse effect' and the build-up of tropospheric ozone and could have potential carcinogenic effects, which increased by 41 per cent.

It was further recorded in the 1993 Bulletin that trends indicated that in a 'do nothing' situation, energy consumption and CO_2 output by the transport sector would likely increase by 24.6 per cent between 1990 and 2000. The worsening traffic congestion was also noted. Between 1970 and 1989, the volume of road traffic in terms of vehicle miles is said to have doubled both for private cars and road haulage. On the basis of trends since 1975 it is claimed that the number of cars would increase by up to

30 per cent during the period 1990 to 2010, a figure which is in general accord with the information contained in Figure 1.1 relating to car ownership in Great Britain.

These figures reflect the situation in the European Union but, as has been said, pollution is worldwide. However, the mix of pollutants for which different nations are responsible varies according to their industrial activities, their way of life and the extent to which they are pollution conscious. For this reason it is not practicable to give any worthwhile or meaningful generalized figures on a global basis of the proportions of the various types of pollutants and their source (ie, from transport, industry, domestic activities, etc) in this, what is only a brief review.

Governments accept the strong majority view that there is global warming and that it is being caused by carbon dioxide and other gases. It was, therefore, agreed at the 1992 United Nations Framework Convention on Climate Change at Rio de Janeiro in Brazil to stabilize emissions of carbon dioxide at 1990 levels by 2000. But such good intentions can only come to nothing unless there is also a willingness to accept some sacrifice. In the event it became unlikely that many countries would meet these voluntary targets and a further conference was convened in 1997 at Kyoto in Japan in an endeavour to establish legally binding commitments to curb emissions of carbon dioxide and other global warming gases such as methane.

Discussions ensued at Kyoto, particularly around the three main powers of the industrialized world, being the European Union, the USA and Japan. But much of the world's pollution emanates from the developing countries, and the USA in particular was conscious of this fact. The main upshot of Kyoto was that a treaty would commit signatory powers to reduce their combined emissions of six key heat-trapping greenhouse gases by at least 5 per cent by the period 2008 to 2012. Emission levels of the three most important gases, being carbon dioxide, methane and nitrous oxide, would be measured against a base year of 1990. A declared number of countries must ratify the Kyoto Treaty if it is to come into effect and this would have to include developed countries in sufficient numbers to account for more than half of the global CO_2 emissions.

Some nations have already signed, including Great Britain on behalf of the EU (the constituent countries of which will then agree between themselves), on the principle that the developed world should give a lead and show the way. But there is another body of opinion that the others must also make their proper contribution to total emissions. The USA has expressed its reservations with the view that there should not be any less binding commitments for the developing countries as they are becoming increasingly responsible for greenhouse gases.

It has been said that this is an ongoing process and at this stage there the matter rests. So far, therefore, nothing positive seems to have emerged in the process of combating global warming and, in any case, the proposals are not exactly awe-inspiring. But if the Kyoto Treaty comes into force it would at least be a first step which might bring matters to the fore, focus attention on to this crucial issue and hopefully become the forerunner of other more positive measures.

POWER AND POLLUTION

Introduction

Reference was made in Chapter 1 to the various forms of motive power that have been in use over the years. This is a follow-on from there as, more recently, various adaptions and alternatives have been introduced which have more environmentally friendly properties. For further consideration, therefore, are the original widely used fuels (being oil-based petrol and diesel) along with their detrimental features in environmental terms, together with the alternative fuels, that are now being either adopted or developed as which are polluting and/or are not reliant on oil.

Following the basic petrol and diesel fuels, there can be no declared 'order of merit' amongst the alternatives. Much depends on local availability and, in any case, this is an area of constant research and development. New data is constantly being obtained which tends to move the degrees of advantage from one to another. The subject of alternative fuels is complex and only an in-depth study would permit an authoritative assessment of any preferred course of action. What follows is intended to give just a general picture of the overall situation.

Petrol

Used mainly in cars, light vans and small boats, emissions from petrol engines include, in varying proportions, carbon monoxide, carbon dioxide, nitrogen oxides and lead compounds. Unleaded petrol has been introduced to deal with the latter but otherwise it still produces all of the other toxic emissions. It is not, therefore, as 'green' as the colour of the petrol pumps might suggest. Catalytic converters are now available which convert most of the remaining exhaust emissions to less toxic substances although they do increase the emissions of carbon dioxide and nitrous oxide. As lead destroys their chemical content, catalytic converters can be used only with unleaded petrol and they are effective only when the engine is warm. In the absence of further development, therefore, they are of no benefit for short journeys when the engines are cold (which is when emissions are at their highest): many motorists like to make short journeys.

Diesel

Diesel is usually associated with the 'heavies' in road transport, being the larger haulage vehicles and buses, but it is used also for many smaller vans, even some cars and also extensively on the railways. Diesel could also be used by ships but most use the heavier and cheaper fuel oil (see below). The combustion process of a diesel engine is more efficient than that of a petrol engine (which makes it more economical), it produces much less carbon monoxide and also less nitrogen oxides and hydrocarbons. It does, however, produce far more particulates. Nevertheless, over the years, diesel has been a favoured alternative because of its economy and very small carbon monoxide factor. But there has been increasing concern over the serious effects of the particulate matter at the same time as the catalytic converters were combating the carbon monoxide issue relative to petrol engines. In health and environmental terms, therefore, the balance of judgement began veering away from diesel and towards petrol. However, the development of a more completely refined (and more expensive) ultra low sulphur fuel (or City Diesel) has significantly reduced the level of sulphur dioxide and particulates. It also eliminates black smoke and hence the dirty atmosphere and smell which is so often associated with busy urban centres due to emissions from diesel engines. Oxidizing catalysts may also be fitted to the exhausts which, as with the petrol engine, convert carbon monoxide and unburned hydrocarbons to less harmful carbon dioxide.

Liquid petroleum gas (LPG)

LPG is a by-product of petro-chemicals and consists mainly of propane and/or butane. It is generally more efficient than either petrol or diesel and it has lower CO and CO_2 emissions. But production is limited without incurring higher energy consumption and emissions during the refining process. In any case, as the depletion of fossil fuel stocks is one of the reasons underlying the exploration of alternative fuels, an extensive switch to LPG (which would not help in this respect) is not anticipated.

Natural gas

Consisting mainly of methane, natural gas comes in two forms, being compressed natural gas (CNG) or liquefied natural gas (LNG). CNG is the more commonly used and is the identical product to that used domestically for cooking and heating purposes.

In transport terms, as the world is better endowed with natural gas than it is with oil, this in itself is a powerful reason to develop the use of this type of fuel. But it is also important in the pollution context as it is more environmentally friendly. It has a cleaner burning engine which means that there is less particulate matter as well as less CO, CO_2 and N_2O than there is with petrol or diesel engines. It is suitable for both light and heavy vehicles but for practical reasons there cannot just be a few isolated users. An extensive distribution network must be in place before motorists are likely to purchase a gas-fuelled car but, on the other hand, suppliers are not likely to invest in specialized filling stations, etc without a ready demand. For a general 'take-off', therefore, there needs to be a planned programme. This could be encouraged by tax incentives (which might run contrary to a tax-encouraged switch away from cars and on to public transport) but such a move has nevertheless already been witnessed in parts of Germany, France and Italy. Otherwise, it is convenient for bus companies and other localized transport operators. It is of note that bus companies across the world from North America to Europe to Australia are switching in whole or in part from diesel to CNG.

Kerosine

A light distillate, kerosine is produced during the refining of crude oil, as is petrol and diesel, and is used by commercial aircraft. But it does pollute to varying extents depending on whether the aircraft is taxiing, taking-off, cruising or landing. Of particular concern in this instance is that the pollutants from kerosine are emitted in the most vulnerable layers of the atmosphere.

Fuel oil

The great bulk of the world's shipping uses fuel oil (bunker oil) for propulsion purposes. It is a refinery residue and is cheap but it is not environmentally friendly. The burning of fuel oil causes the emission of quantities of CO, CO_2, SO_2, NO_x and VOCs and there is therefore a case to consider possible alternatives.

Alcohol

Two forms of alcohol may be used for transport purposes, being methanol and ethanol. Methanol is produced from natural gas by processing oil, coal or decaying vegetation whilst ethanol is distilled from natural sugars and starches. Both can be used as an additive to other fuels (such as petrol and diesel). Their use gives a better performance when used with petrol but not when used with diesel but there are lower emissions other than NO_x, particularly from methanol which is produced from wood. However, there are disadvantages such as cost, availability, corrosion, etc which gives weight to the thinking that there will not necessarily be a long-term future for these biofuels.

Hydrogen (H)

As any student with an elementary knowledge of chemistry will know, hydrogen is a constituent part of water (H_2O). It is the cleanest fuel as, when burnt, it produces only water with maybe an element of NO_x and there is no chance of it becoming scarce. On the face of it, therefore, hydrogen is the ideal fuel but unfortunately it is not an element that is available in its natural state. That is where the problem lies and is the subject of current research.

It is an expensive fuel in that energy must be used to separate it from oxygen (maybe by electrolysis of water) only to burn it to get back that same energy (which is then sufficient to propel rockets into space). It must also be stored and distributed in safety. Environmentally, it would be superior to kerosine for use by aircraft but it is an explosive gas and both its storage and use also remain the subject of research (and remember the airship disasters of the 1930s where the craft concerned were filled with hydrogen, albeit for lifting purposes).

Research and experiments are ongoing and in the longer term, for both availability and environmental reasons, the use of hydrogen as a fuel for transport purposes may have promise.

Electricity

Preamble

Having now reached electricity in this review of power in relation to pollution and future availability, thoughts revert back to Chapter 1. It was there noted that whilst electricity is produced at a remote source and fed to fixed tracks, storage on a free-moving vehicle in the form of a battery still remains the subject of research. There are therefore two separate issues and they will be dealt with as such but it is only the generating process that has environmental connotations. The actual use of a battery-powered vehicle on the road produces no emissions.

Generation

Heat converts water to steam which drives a turbo-generator to produce electricity. Alternatively, direct water or wind power can be used for the same purpose. That in simple terms is the basic principle of a power station. The fundamental issue in the context of atmospheric pollution and the continued availability of oil is that if water is to be turned into steam, what fuel is to be used? The possible options to heat water are:

- coal
- oil
- natural gas
- refuse
- solar
- nuclear fission

Then there are the natural and renewable (pollution-free) 'direct drives' created by the flow of water emanating from rivers, tides and waves and by the wind.

The natural and renewable pollution-free sources of power (which also includes solar heat) are, of course, the ideal but can be used only where they are readily available and where their power can be suitably and economically adapted. In any case, for large-scale power-station requirements it is likely that the main natural source will be the flow of water from rivers (to produce hydro-electric power). Even so, active consideration is now being given to the establishment of off-shore turbines. If this came to pass then some coastal areas could become dotted with 200 feet high masts equipped with revolving blades capable of producing electric power – not cheap, but environmentally friendly. The Government of Denmark, for example, has launched such an investigation and it has been suggested that a 10 per cent reduction in the use of fossil fuels for this purpose could be met in this way. It has been further propounded that Denmark might possibly produce 40 per cent of its electricity requirements from wind sources by 2030. Otherwise, it is back to the conventional fuels which have already been considered in relation to their pollutant properties and continued availability.

Coal is a major source of energy. It provided power for the Industrial Revolution in Great Britain and today it still plays its part in the production of electricity. But coal-burning power stations release CO_2, SO_2 and NO_x into the atmosphere which contribute to both the greenhouse effect and acid rain. Because of this, many coal-fired power stations have been converted to either oil or natural gas. Oil has an advantage over coal in this respect, but still better in terms of atmospheric pollution is natural gas. However, it is at this point where thoughts turn to nuclear power.

The principle of a nuclear power station is the same as that of a coal or gas-fired power station. The major difference lies in the fuel that is used to boil the water to create the steam to drive the turbine. A nuclear power station relies on a controlled chain reaction among atoms of uranium and it is the splitting of the atoms that is known as nuclear fission. Although supplies of uranium are finite, it is a very concentrated fuel (one small pellet about an inch long is roughly equivalent to 1¼ tonnes of coal) and it can be (and is) recycled which means that this fuel will be available well into the foreseeable future and beyond. Nuclear power does not create significant amounts of carbon dioxide, sulphur dioxide or nitrogen oxides which are linked to the greenhouse effect or acid rain, and to this extent it is environmentally friendly. However, the making of nuclear power does create radioactivity which can be dangerous to humans, animals and plant life if not properly treated and managed. Nevertheless, subject to proper controls, it is claimed to be safe and it has the potential to become a massive source of energy for electricity.

This, then, is a brief summary of the benefits and disbenefits of the various types of electricity-generating stations. Note again that the degree of atmospheric pollution attributed to electric trains, trams and trolleybuses (and by road vehicles if and when the battery is perfected) is governed by the fuel that is used (if any) in the power stations. Many of these systems are not, therefore, pollution-free.

Batteries

There would now seem to be little doubt that the longer term future of road transport lies in the battery or battery-assisted (dual mode) form of propulsion. Advanced batteries such as lithium ion and nickel metal show promise. If and when the battery car is perfected and becomes widely used then, in environmental terms, it would be superior in every respect. But that is a part of the technology of transport – not part of the basic principles of transport.

Overview

It should be noted that the different fuels that are available have been discussed in this sub-section only from an environmental aspect. Some fuels may be more economic than others and there may be differing maintenance and cleaning costs. Furthermore, the actual fuels which are adopted are influenced by natural availability country by country. Account has been taken of the fuel only when it is in use, as allowances have not been made for the different impacts on the atmosphere made during the course of production or distribution.

In a brief summary such as this it is not practicable to quantify the net advantages of any one type of fuel. An advantage one way can be counterbalanced by a disadvantage in another; the optimum will depend on circumstances prevailing at the point of use. But the foregoing sections have given a basic picture of the properties of the different fuels from the environmental angle. The information on the alternatives is also relevant in the context of a future scarcity of oil.

Also important is sustainability, being something to which reference is constantly being made in environmental terms. To quote an example, it will be recalled from Chapter 5 that one of the key strategies of the State Transit Authority of New South Wales was to ensure that development is ecologically sustainable. But its meaning is not necessarily clear. Sustainable development was, in fact, defined in 1987 by the World Commission on Environment and Development as development that meets the needs of the present without compromising the ability of future generations to meet their own needs. Most importantly in this context is the avoidance as far as is possible of the use of finite resources of fuel and other essential materials. Another important factor is the contribution to irreversible climate changes through the emission of greenhouse gases. Both of these factors have, of course, strong repercussions on transport; the extent to which it is provided, the fuels that are used and possible switches between the modes. To sustain future quality of life, therefore, policies must as far as possible be geared to using only those resources that are replaceable. Again to give examples, water and wind power will continue to be available to generate electricity regardless of how much they are used. Uranium will be available for a very long time but oil will not. It has been said that wood is a valuable commodity but as trees are felled, an ecologically sustainable policy demands that new trees are planted in their place.

NINE

A Pattern for the Future

PREAMBLE

The demands for transport are ever changing in what is an ever-changing world, as are the means of meeting those demands. Vehicles, vessels and aircraft continue to become larger, faster and more complex. The advent of the computer and electronics has revolutionized previous practices in everything from administration to navigation; from advance booking to fare collection; from manual operation to automation. Manpower is giving way to machine. Under normal circumstances, therefore, a look into the future would involve a consideration of how new demands can be met with the sophisticated tools that have become available. More and more people wish to travel, particularly for pleasure, and an increasing range of facilities is encouraging them to do so. With a ready demand and state-of-the-art equipment to hand, this could be an exciting time for the transport industry, particularly for those who cater for the longer distance and international traveller where there is still much reliance on public transport. Unfortunately, however, all is not well. As has been said so many times before, for the shorter and local journeys, most people no longer require public transport. They want their own private conveyances, which has made the provision of public transport for this particular sector difficult. But more than that and something to which attention has been drawn throughout this book; there is now also the problem of space (on the road) and atmospheric pollution. The same problem is now spilling over to air travel with its cheap package holidays. Lack of space (among other things) is now becoming an obstacle to the provision of adequate airport capacity, approach roads and aircraft take-off and landing facilities. These features are fundamental when considering a pattern for the future.

In previous editions of this work it has been the practice to devote the final chapter to the future. It contained a few pages which portrayed the views of the writer on what might happen in the world of transport in the years to come. Speed and size were favourite topics, with Concorde and 'jumbo jets' in the air, large hovercraft on the sea and advanced passenger trains very much in the news. Faster and bigger was then the order of the day as it still is. Then there was the predicted shortage of oil with warnings of an imminent danger of wells running dry. If this was to be the case then electric vehicles would become all-important. More research was necessary to perfect battery-powered cars (it still is); trams might return to their former glory and alternative fuels would be necessary for power stations. It was not until the 1990s that thoughts began to turn more seriously to road space running out and also to the sombre warnings about pollution and global warming.

In previous editions this was not only the last chapter, it was also the shortest and probably the least important in terms of educational value. But planning for the future is now more important than it ever was and greater consideration must therefore now be given to this aspect. A study of transport, even in general terms, would not be complete without it.

It is accepted that the environment is an all-important issue and one that will not go away. However, the very act of living has detrimental effects and as people must live, its preservation becomes a matter of judgement and degree. Even so, the lack of road space (and air space) with resultant pockets of chronic congestion is something that will have to be reckoned with, requiring as it does immediate and drastic measures. Furthermore, although oil wells are no longer thought to be in immediate peril of exhaustion, supplies are still, nevertheless, finite. To summarize, therefore, the major transport issues requiring attention may be identified as:

- space (in which to move around)
- the environment (with particular reference to pollution)
- oil (and alternative fuels)

In a car-orientated society, the imposition of demands such as full public transport services at all times with cheap fares can only spell despair to operators. Carrying around empty seats is not a profitable venture. But if the people were divorced from their cars, at least for their local trips, the demand situation would be different. The buses and trams would then find their niche and monetary support should not then need to be as heavy. Furthermore, with the absence of cars and hence congestion, operating costs would be less and the public requirements referred to in Chapter 6 would be more easily met.

It must now follow that a pattern for the future cannot involve just a plan for public transport and road construction geared to predicted demands. The environment has become a crucial issue and lip service is already being paid to the environmental damage that is being inflicted on the world (in this context by transport) to which the palliatives that have already been introduced bear witness. Pressure groups and others have alighted on their favourite targets, being private cars and the heavy trucks. But just to list their disbenefits in terms of pollution, congestion, health hazards, global warming and the like will not cajole motorists out of their cars and on to public transport, which is understandable. Alternatives, if there are to be any, must be adequate and (under the circumstances) acceptable and they must be seen to be such.

GATHERING THE THREADS

We are on a crash course whichever way the statement might be interpreted. We have got to learn quickly (and then do something about it) and if we do not we are going to crash; 'it', of course, being the problems of space and pollution. The time is now ripe, therefore, to sit back, take a deep breath, draw together all of the threads which have become so badly frayed and restrand them to create adequate national lifelines.

Consider now:

- What enables the transport industry to function;
- What damage is it causing in the process;
- What are the various courses of action that would remedy the defects;
- What machinery is available to enable that to be done;
- What will be the available sources of fuel in the longer term.

All of these aspects have been discussed in basic terms in the previous chapters. It is necessary now to bring forward that information and without any further cross-references develop a strategy for the future. But before proceeding further, remember that whilst everything that has been said so far represents fact which must be read and digested, the interpretation of those facts in order to create a sustainable system for the future involves subjective thought. From this point onwards, therefore, what is said represents one possible course of action. The problems are beyond dispute but how they are solved is matter for debate. Furthermore, what follows does not represent a suggested master plan for any one place or, indeed, for any one country or for implementation all at the same time. It is something that might be introduced by degrees as and when circumstances are appropriate. It is put forward as just one line of thinking with the principles involved available for consideration. Some of the assertions that follow run contrary to views expressed by the European Commission in its proposed 'Citizens' Network' as referred to in Chapter 1 and elsewhere. But students of transport should read all of the proposals put forward by different bodies and come to their own conclusions.

Before proceeding further, however, there is one aspect that can be disposed of at the outset. It is mooted in some quarters that there should be more walking and that many of the short trips that are made today could be done in that way as it is pollution-free. But much depends on what is meant by 'short' and in any case some people use their cars (or any other form of transport) because they do not consider it is safe to walk, particularly at night. Walking is a pleasant and healthy recreation in the right place at the right time. But in the opinion of the writer, walking does not blend easily, other than for short stretches, with a day's work. People did it in the eighteenth century but we could be excused for expecting something better today. It is not really tackling the problem and to the writer it seems almost ludicrous to classify walking as an alternative mode of urban transport. Walking, therefore, other than for an accepted distance to and from the nearest bus stop, will not form a part of the proposals. Furthermore, if people on bicycles continue to constitute a source of danger in the ways that are illustrated, it will only be a matter of time before there are demands for cycle restraints.

OWNERSHIP AND CONTROL

When considering a pattern for the future, the vexed question of public or private sector ownership cannot really be ignored. It is a topical subject. Indeed, politicians (worldwide) are telling us that the current swing to privatization, aided and abetted by the free forces of competition, is achieving greater economy and better facilities. The question now, therefore, is whether this ideal should be taken on board as the framework for the pattern that is now being proposed.

There is a fundamental difference in the conveyance of goods and passengers. Goods transport is more 'tailor-made' to the requirements of the users and the service (or 'wagon space') is not as 'perishable' in the way that a passenger seat is. This can reduce the degree of both cost and waste in the carriage of merchandise compared with passengers and renders a competitive environment more appropriate. But if public passenger transport is to be adequate in that its facilities are available to meet all reasonable demands with attractive fares, then it is far from clear how its provision could be left just to free market forces. This is something which is important in this context and also in the financial context which will be considered below. But in the case of goods transport, ownership would not appear to be a crucial issue.

At this stage, control must also enter the equation. Privatization is one thing but control is another. It is submitted that regardless of any private or public sector allegiance, in no way must there be a surrender of control; this means quantity control. In everything that follows it is assumed that quality controls will remain and be strictly enforced. But to return to quantity control: the writer is of the opinion that in countries where there has been deregulation then there must be reregulation. Its presence is crucial for:

- local bus services, be they urban or rural, to ensure the provision of an 'adequate' integrated network with reasonable fares, to co-ordinate with any parallel rail or local rapid transit services and to eliminate waste. Unnecessary journeys means unnecessary pollution;
- longer distance coach services if it became desirable to transfer traffic to rail in the interests of road space and/or the environment;
- fares on rail services in urban areas where they make a contribution to local passenger networks; in rural areas where they provide a valuable 'lifeline' and to offer a reasonable alternative to any coach services withdrawn;
- safeguard against any substantial reduction or complete withdrawal of any rail service;
- road haulage services (both 'hire and reward' and 'own account'), if it became desirable and practicable to transfer trunk hauls from road to rail, inland waterway or coastal shipping in the interests of road space and/or the environment;
- air charter services, sub-standard fares on scheduled airlines and the fare content of package holidays if it became necessary to restrict operations in the interests of space and/or the environment.

The mechanics of such controls will not be described here as a study of modal transport operations and legal issues will reveal various methods administered by government-appointed bodies. The actual regulations are in any case likely to vary country by country.

Having established the view that a degree of regulation is desirable, and certainly on the passenger side, attention can now revert to the ownership aspect. It will be seen from the above that considering all of the modes, the most widespread controls need to be applied to local bus services. This does not necessarily involve large-scale operating units although the writer sees no need for any restriction in this respect. In practice there might well be large units. Certainly a single operator with a single management within a conurbation would achieve the co-ordination requirement and suitable controls would avoid the inherent dangers of a monopoly, but large-scale administration does not always mean maximum economy. The results could be achieved equally well with smaller companies working to a route tendering system controlled by a professional overall regulating authority. The smaller units would be privately owned but if the large undertakings came (or remained), the writer would not be overconcerned whether the stockholders were part of the public or private sector, providing those in day-to-day charge were professional and not political.

ROAD SPACE AND THE PRIVATE CAR

In many urban areas road space is rapidly running out. This situation is inextricably linked with the private car and the resultant chronic traffic congestion and atmospheric pollution is the subject of much debate. Car restraints, therefore, are a crucial part of any pattern for the future but before we heed the advice of some pressure groups and the anti-car lobby, it is first necessary to consider whether the car does have any real place in meeting national transport needs.

The beneficial features of the private car may be summarized as follows:

- It is the most convenient of any mode both for the journey and for carrying shopping, luggage, etc.
- It is something that everybody wants and provides a major form of recreation.
- Some motorists claim it is a crucial element of their everyday life (but maybe only because they have built their habits and their personal arrangements around it).
- For people living in isolated rural areas it is often their only link with the outside world. Conversely, it facilitates a tourist invasion of their homelands if they are also areas of natural beauty and this has a damaging effect.
- Depending on the number of passengers and extent of use, under some circumstances its costs compare favourably with public transport. For 'extra' journeys, the marginal costs of the car are cheaper.
- It may be the cause of congestion now but it did not originate it. Congestion and haphazard vehicle movements existed in the days of the horse, albeit in smaller and more compact areas.
- It may now be a major contributor to global warming but the writer submits that global warming (if it is to come) would have come in any case (see Chapter 8).

- Use of the car for travel in the peaks to and from work relieves operators of the diseconomies of maintaining a fleet of low-mileage peak-hour-only vehicles.
- If all motorists suddenly abandoned their cars and turned to public transport then some light and heavy rail transit systems could not accommodate the extra influx of traffic.

Clearly, the car cannot be restrained lightly. On the other hand, with less than, on average, two people per trip, it is extravagant in its use of road space and in some places its ever-increasing numbers can no longer be accommodated. It also causes a high degree of atmospheric pollution. Furthermore, its parking habits (ie, on the roadside at the point of destination and even also for residential purposes) are not acceptable as it is parked cars that block the highway and materially add to congestion. Whilst, therefore, it would be unrealistic and unreasonable to propose total bans as such, consideration must be given to arrangements necessary to reduce their numbers.

An alternative approach in respect of roads and traffic congestion is to give more space by the construction of new roads. This aspect will be discussed in the 'Land use and planning control' section below.

PARKING RESTRAINTS

A means of controlling both the use and the ownership of cars would be to institute a system of parking restraints, together with a gradual introduction of a ban on on-street parking for residential purposes. The idea is not new. In parts of Japan a certificate that a place is available to park off the highway must be produced before a car can be licensed. As a start, there could be a parking ban on one side of the road only on alternate days as in France. A public highway is a traffic artery along which people travel to wherever they want to go. The practice of blocking it for use as a free parking lot, whilst other motorists who provide their own off-street garages pay for the privilege in the form of rates or rent, has no logic. It would be apposite, therefore, to introduce controls on a selective basis as and when the needs arise. At destination points there could with advantage be a ban on parking and even on short-term waiting for any purpose along bus routes and other roads which have a heavy traffic movement. This would be without exception and no special privileges would be afforded to any particular categories of people. It is worthy of note that in many instances, such stopping is already prohibited but the law is disregarded. In these cases it becomes a matter of enforcement. Heavy penalties for those involved plus withdrawal of driving licences would make a yet further contribution to a reduction in the number of cars on the road.

Within the suggested plan, provision of public car parks, free or otherwise, on-street or off-street, would be prohibited within congested zones according to circumstances and as may be determined. Any existing private off-street parking places (business or domestic) could remain and similar facilities would likely be afforded to new construction. Applications by businesses for limited off-street parking spaces that are also available to the public on a strictly short-term customer only basis would be considered on their merits. A bank, for example, might well be successful in this respect, whereas a food outlet would not. Parking facilities for staff would be severely restricted. Car parks would, however, be provided outside congested areas for use (at cost) in conjunction with park-and-ride facilities if there was sufficient demand, but with better and cheaper public transport that demand might not be high.

The imposition of parking restraints with supporting alternative facilities would be administered through the system of planning controls which will be considered later in this chapter.

A PRICING MECHANISM

A more subtle (and could be parallel) method of discouraging use of private cars in urban areas would be through a price mechanism. The arrangement in most countries is that the motorist first bears the capital cost of purchasing his vehicle. He then has to tax and insure it. His only expenditure once on the road is for fuel (which can also involve tax to varying degrees) plus a small amount for maintenance (which he probably forgets about at the time). In other words, the standing costs are substantial. On the other hand, travel by public transport is a pay as you go process (unless a prepaid season ticket for regular users is available which does then also make it cheaper). But having met the standing charges of a car, the motorist will not be happy to pay the full public transport fare for an extra trip compared with only the marginal cost involved by driving himself. It would be unreasonable to expect him to do so. The situation must therefore be reversed with motoring costs put on a mileage basis and public transport fares (for journeys within urban areas) paid for in advance as an 'overhead' cost of general living (at a relatively cheap price).

Although many countries already add tax to the cost of fuel, it would be necessary to receive all of the desired revenue through tax at the sales point. However, this is the easy part. Some countries already do that and it does at the same time go some way to transfer the costs from overheads to mileage. But it does not go all of the way. Insurance is also an item and it can be a heavy one. Most countries have a system of compulsory insurance (it is submitted that they all should) but there is a range of premiums depending on the extent of the insurance and the status, place of residence, age, bonuses, etc of the insured. There are also many different insurance companies. To achieve the purpose, the insurance element would also need to be added to the cost of fuel but this does involve a new and complex problem necessitating research. Even so, the electronics industry might now come to the rescue. Maybe, instead of paying an annual premium, the insured person could be issued with a plastic card with a code number rather like the British Inland Revenue has for the payment of income tax on the pay-as-you-earn basis. This card would be presented at a petrol station where payment would then consist of cost of fuel plus tax plus insurance, the latter being variable. There could also be two or more levels of tax according to where the petrol stations are situated, be they in busy urban areas (where it would be high – even very high) or in a country town or rural area (where it could be lower). It would be geared to put the overall cost of fuel well above that of public transport or even to part-subsidize public transport.

With this arrangement it is suggested that the need to impose tolls on selected roads or motorways would not arise and would be applied only on specific short sections (eg, bridges, tunnels and the like) at discretion and for special reasons. To avoid further complications, all cars would be subject to the same level of taxation and the writer does not in any case see a need to impose heavier taxes on cars that have greater pollutant properties. There have been suggestions to the contrary. The slogan 'the polluter pays', first at Rio and then followed up at the EU, is not new. But it is submitted that to pay for polluting does in a sense afford an entitlement to pollute. But people should not be able to pay to pollute – they should not pollute at all (at least, not more than is necessitated by the best designed engines and fuel). To return to the payment scheme, safeguards would be necessary to ensure that the customer produced the correct card when buying petrol and if no card was presented then a special (high) charge for the insurance element would apply. Certainly liaison would be necessary with insurance companies and appropriate software developed. Whilst the ultimate system would not necessarily be exactly along the lines suggested, the principle is there and its implementation in some form would be crucial to the scheme.

As a brief digression, it is of note in this context that, at one time, owners of motor vehicles in New Zealand paid their necessary third party insurance premiums to the 'Deputy Registrar' along with their vehicle licence fee, although this particular piece of legislation has now been repealed following the introduction of an Accident Compensation scheme. Speaking of New Zealand, however, it is also of note that there is here an example of the tax which is levied on petrol actually having a part dedicated to road construction and maintenance, but even so this reflects only an average cost and there remains concern in New Zealand that there is no direct relationship between road pricing and actual road use. No account is taken of, for example, vehicle weight or the time of day (and hence traffic congestion) and this has a strong bearing on costs. Accordingly, the Government appointed a Roading Advisory Group in 1997 to provide proposals to ensure a safe, sustainable, fair and efficient road system at reasonable cost.[1] The resultant proposals embodied a comprehensive review of road ownership, finance and administration. Pertinent to the topic under review, however, there is a recognized need to relieve traffic congestion; hence the concern about payment according to the time of day. The point is that in a commercially structured road system, pricing can be used to reduce future demand. It is estimated by Auckland Regional Council that if average pricing continued, NZ$2 billion worth of new roads would be needed in Auckland by 2021. With marginal congestion pricing it is thought that this figure would be reduced by 90 per cent. The proposal, therefore, is that there should be a gradual transition to this end as new and more cost effective charging technology for direct road pricing (or 'value pricing') is developed. The principle is similar to that of a telephone call, the cost of which varies according to the time of day. In fact, some New Zealand trucking firms are already using this technology to manage their vehicle fleets and in the course of time its extension to private cars should become practicable.

Returning now to the main narrative and the world at large, whereas the cost of motoring has been put on a mileage basis, the cost of travel by public transport would become a prepaid 'overhead' expense. Travel cards valid for unlimited travel within defined areas and for defined periods would be available at reasonable cost. Certainly, this type of ticket is already available in many places. The difference is that with costs on a different basis and with more attractive and comfortable services (and less delay due to traffic congestion), more people would be encouraged onto public transport and hence benefit from the low (nil) cost of any extra trips.

This proposal does give rise to one further observation. There have in the past been politically motivated suggestions to allow free travel for all on public transport in urban areas, the theory being that it would be a move to encourage people to get out of their cars and on to buses. The writer sees this as an undesirable step that would not achieve its purpose for the following reasons:

- It would leave the cost of private motoring unchanged and therefore at a level that the people are prepared to pay.
- It would make bus service provision difficult to plan. With costs borne by taxation everybody would be entitled to a service which could not

be 'rationed' as can, for example, the number of library books that are issued or the number of times that household refuse is collected.

- It encourages unnecessary travel, particularly for very short trips.
- It is pointless to charge (or offer free for all) less than what the traffic can bear (see Chapter 6).

The proposal to transfer the transport 'overheads' from cars to public transport overcomes the first of the above objections to 'free fares'. There are, nevertheless, still elements of the second and third issues contained in any proposal for cheap travel card type tickets. However, bearing in mind all the circumstances appertaining to the scheme as a whole, it is submitted that it might be beneficial to accept the remaining, now relatively minor, deficiencies.

OTHER ROAD TRANSPORT

So far, the problem of traffic congestion in urban areas has been dealt with by an endeavour to shift motorists from private cars to public transport through parking restraints and the price mechanism. Nevertheless, the problem will not have been cured. Goods vehicles and taxis could still remain an obstacle to free-flowing traffic and selective restraints might have to be applied, remembering always that shops cannot live without deliveries of their wares. Also on the positive side, taxis complement the fixed routes of public transport and are convenient, particularly for those encumbered with luggage, etc. But, like the car, they are inefficient in the use of road space (without the driver their average load is less than one), through sheer weight of numbers they can cause congestion, they delay buses and are heavy pollutants in relation to the numbers of passengers involved.

Restraints might have to be applied to both of these classes of traffic and arrangements for shop deliveries at the rear of the premises will need to be the policy for the future. Where not immediately practicable, there might have to be restricted times for shop deliveries.

AN URBAN PASSENGER TRANSPORT SYSTEM

Setting the cost of motoring on a mileage basis is half the exercise. The other half is to put in place adequate public transport alternatives (and what follows would of course apply equally if harsher forms of car restraints were imposed). For large urban areas this would involve whatever modes are appropriate, ie, diesel buses, dual-mode diesel/electric buses and/or trolleybuses including guided systems, trams, light rail and heavy rapid transit plus the utilization of suburban railway systems where they exist. All would need to be examined on their merits which would involve cost in relation to potential traffic and atmospheric pollution.

Remember that the establishment of a public transport alternative to the car requires consideration of:

- the maximum distance any one passenger could be expected to walk to a bus stop;
- unless using an infrequent service and aiming for a specific journey, the time that any one person could be expected to wait.

With that criteria established, a network would evolve and timetables be prepared. The services would need to be reliable, the cost reasonable and the ride comfortable. Much play is made of a truly integrated seamless system whereby through journeys are facilitated with one ticket and everything connects with everything else. That is certainly the goal of the European Commission, as was indicated in Chapter 1. But in practical terms the writer submits that no system can ever be as fine-tuned as that. In any case, passengers have different priorities. Whilst some may want a connection, others are more interested in starting times at work, etc. Furthermore, the mechanics of the timetables with running times in relation to what should not be excessive stand times could render perfect connections everywhere impracticable if only in mathematical terms. To go on then to gearing such a perfect network to connections with other longer distance services and even other modes would make the whole thing an impossibility even on paper, without even progressing to the practical problems of delays on the road. The writer considers that the most that can be done is to bear in mind all of the requirements and produce an optimum network. The systems concept in transport management (also discussed in Chapter 1) is one thing but 'integration', as seen by politicians, is quite another.

If the traffic congestion factor is overcome, then reliability is eased and it then becomes a question of proper management. The availability of cheap fares will involve monetary support from public funds and that aspect is commented upon under the finance section below. The comfort factor is something where there is now much scope for improvement and which calls for specific comment. However, as this is a book about the principles, the subject can be dealt with in only general terms.

Over recent years there has been a swing to mini and midi buses which have become a panacea for all ills. Whilst it is accepted that there is no sense in carrying around surplus capacity, at least for some journeys they become grossly overloaded with the result that passengers are forced to stand in most uncomfortable conditions. Even some double-deck buses have been replaced by standee type single-decks with the same effect.

Then there are the low-floor single-decks which are unpleasant to ride in and invariably involve a dangerous step halfway along the interior of the vehicle. All this is done for the sake of the elderly and infirm. However, it will now have to be understood that travel of this kind is not good enough for the displaced motorist. It will also have to be realized that if these new circumstances come (and surely they must come) then there will be a new influx of passengers. They will be the working population, the commuters, the younger and middle-aged groups – the mainstay of the country; and they will want (and deserve) to travel in comfort. They will want a seat, certainly on road vehicles. Furthermore, even small buses are extravagant in their use of road space and are self-congesting. The bus industry will have to do better than that. It will have to give the majority and the essential riders priority treatment and an acceptable service as the elderly and the infirm become the minority users. Once the traffic is there, the standard-type vehicles will need to revert to high capacity double-decks with whatever form of propulsion is selected. The only exceptions would be on those routes where physical limitations or paucity of traffic call for something smaller. The writer submits that motorists displaced from their cars would not appreciate the sight of myriads of little buses cluttering the highway and distributing unnecessary pollution into the atmosphere, being something for which they have already been castigated.

These are the possible lines along which urban passenger transport systems might be revitalized. In practice it is unlikely that the same thoughts will be applied to rural areas for a restoration of the erstwhile country bus networks, or at least not for a very long time. Wherever the car can be contained within the existing highway network and there are no other physical grounds on which to exclude it, then the wishes of the people should be respected and it would stay until circumstances dictate otherwise. This will be a disappointment to the bus lobby as it will to those in rural areas who have no access to a car. Once again, the minority have to suffer although any existing arrangements to support rural bus services would continue and could be increased to the extent that funds might permit. However, if the suggested fiscal arrangements came and public transport became more popular then even the rural services might benefit. Quantity controls would still apply to eliminate waste and maybe something better could be provided out of the 'fat' of a neighbouring urban system.

RAILWAYS

Worldwide, railways today range from dilapidated, outdated, underused assets through to high-speed lines forming national networks with all modern comforts and conveniences and providing a crucial input to total transport systems. This proposed pattern for the future would incorporate railways even though less costly alternatives are sometimes available. Railways would be retained and developed where appropriate, not only for pure transport purposes but also because of lack of space on parallel roads and for environmental reasons. The Institute for European Environmental Policy is on record[2] as observing that both France and Germany had at one time a policy of defending their railways from competition by road and imposed restrictions on lorry movements. However, although this may have been environmentally beneficial it was not in accord with the EU Common Transport Policy and things are now different!

It is not appropriate to discuss specifics here but on the passenger side, main intercity routes would be developed wherever appropriate, as would suburban and urban rapid transit systems. On the freight side there is scope for development and modernization as necessary, not maybe in economic terms but again in the interests of parallel road space and the environment. The possibility of the trunk haul of at least some merchandise being taken off the road and put on to rail would be pursued; such a policy would figure in this future plan.

LAND USE AND PLANNING CONTROL

Land use has long had a bearing on transport in that transport has had to meet the resultant demand. Transport also has a bearing on land use in that it has created demand and land use has been shaped accordingly. In this future pattern, town and country planning is considered in the context of land use as it affects transport provision and the construction of transport arteries and outlets in the shape of roads, car parks, railways and airports. It is something that can be controlled through planning permission machinery.

The arrangements that have been suggested so far (ie, car restraints and a pricing mechanism) will have catered for the commuter but the question remains – what about the shopper? It could be said that with cars mostly gone from the worst congested areas, buses will have no problem in accessing town centres and the shops there will blossom. However, although supermarkets are increasingly implementing delivery services, otherwise, in the absence of household door-to-door home deliveries, weekly shopping expeditions with trolleys full of heavy items of food (including maybe frozen purchases) do not really lend themselves to conveyance by public transport (or bicycles or walking). On the other hand, with nowhere to park, the car would no longer be an available option.

The advantages of the private car have already been summarized and the point has been made that in view of its popularity it must remain a part of the transport scene wherever it can be accommodated. It has not been banned as such but in this fanciful futuristic pattern, its use for commuting and shopping purposes to and from congested areas has been all but eliminated. Motorists would still have their cars, even for travel to restricted zones for visits (maybe to places such as banks, schools, travel agents, etc), subject always to provision having been made for off-street parking. They could also drive through them *en route* to somewhere else or, of course, for any other trips outside the zone.

But the needs of shoppers on their main food shopping expeditions have still not been considered. However, there does now seem to be only one option and this is where land use planning and controls come in. The answer can only lie in developing the much maligned periphery or out-town shopping malls, including supermarkets, with large free car parks. Many countries in both the developed and the developing world have adopted such a policy with success and this is what is proposed here as the pattern for the future. Some environmentalists will applaud the suggestion as a means of preserving the character of the older towns (except that they will not wish the countryside to be scarred by shopping development). Other environmentalists along with local chambers of trade will deplore such a move on the grounds that town centres along with the shops there will die. But that should not be the case. Most functions would remain, such as the Post Office, banks, building societies, estate agents, solicitors, travel agents, tailors and other clothing stores, hardware, hairdressers – the list could go on – to which travel by public transport would be perfectly feasible. There would also be outlets for the sale of furniture and other bulky items that would be delivered anyway, plus smaller supermarkets and other food shops suitable for those with just a shopping basket. No – the centres would not die although certainly some 'corner shops' might. But would they? With more expensive motoring and severe restraints on residential parking, some people would find themselves without a car. The local 'corner shops' might well flourish! The establishments in the centre could, of course, duplicate themselves in the outside malls if they so wished.

Also involved in the land use and planning process is the construction of other transport-related infrastructure such as roads, light rail transit, railways and airports. If there is any serious intention of diverting freight away from road and on to rail – and it is suggested that there should be – then suitable in-town terminal sites will be needed which are connected to both road and rail. It is important, therefore, that the planning control machinery ensures that suitable land is not instead developed by non-transport related interests.

Over the years it has been the policy of most countries to introduce traffic engineering schemes and construct by-passes, new roads and full length motorways to relieve congestion and cater for additional traffic. But as it is a worldwide phenomenon that no sooner is new road space available than it becomes filled with vehicles with any relief from congestion being quickly dissipated, it is suggested that as a general policy, there should be no more road construction. Nevertheless, new works including road improvements and upgrading would continue at discretion in the interests of improved layouts, relief of delays at particular pinch points, safety, etc, or any other special circumstances as may be determined on their merits.

All of these aspects would be considered for approval through planning control machinery.

As was the case with parking restraints and land use planning, new main line railway routes (ie, beyond the in-town freight terminals just mentioned) are also heavily involved with landtake and hence planning controls. Apart from the occasional noise and visual intrusion, the railway is reasonably environmentally friendly and its ability to move the masses should be properly exploited. Its main weakness, being cost, is not relevant to this particular section. On the face of it, therefore, it would seem to be something that should be encouraged without being frustrated by planning regulations. However, when it comes to new lines, the construction of a railway inevitably involves encroachment on to other people's land and property, and however much it might be desirable in transport terms, the planning control machinery must be there to safeguard all interests. The problem is accentuated in countries which are relatively small but with a high population density where any new construction is likely to affect land owners and property values. An example is the construction of new high-speed lines to serve the Channel Tunnel which links England with France. In France, large tracts of open country with relatively low population densities enabled ways through to be found for new lines down to the south, etc, but the south-east corner of England, being more developed and compact, produced considerable public resistance to all possible options. Taking the two countries as a whole, the population density of England is 247 per sq km whilst the corresponding figure for France is only 106.[3] In purely transport interests and with the blinkers on, some might say that the bulldozers should come and a way through for the railway found. However, in this pattern for the future, the writer suggests that even in the railway context, planning controls (probably in the form of parliamentary authority) must remain. The same consideration would apply to light rail transit.

AIRPORTS

Another facet of transport which involves capital expenditure, landtake and planning control is the provision and extension of airports which, like the railway, can be influenced by the geographical characteristics of a country. But a busy airport with adequate approaches on the land side and one which is capable of accepting the largest types of aircraft is extravagant in its use of land. Planning consent problems therefore arise which can be as difficult to resolve as are those of a proposed new railway. The question inherent in this pattern for the future, therefore, is where the priorities lie.

It is likely that many, or the vast majority, or maybe even all of the air services which use the larger airports of the world, are international and probably inter-global. If that is the case, then for economic and prestige reasons, those services will be of supreme importance to the nation concerned. No country lives in isolation and adequate transport links with the outside world are crucial and must be maintained. But to be satisfactory, the airport must be sited at a place which is reasonably convenient for the town which it purports to serve, with suitable road and rail links capable of accepting heavy traffic. The airport facilities must be adequate for the throughput of aircraft and passengers and with facilities for all of the many support services. This all adds up to a lot of land being sterilized for any other use. That, then, is the backdrop to which planning applications must be considered and the objections from residents in the shape of noise intrusion at best and blighted property and housing displacement at worst must be heard.

As was done in the case of the railway, a parallel can again be drawn between England and France. France has recently completed work on a modernized and enlarged Charles de Gaulle international airport to serve Paris. A similar enlargement is necessary at Heathrow which is one of (and the largest of) the international airports that serve London. This would be accomplished by the construction of a fifth terminal building. Consent is sought for this work to proceed but resistance from other interests is strong and therein lies the dilemma.

Some of the air traffic which uses the three London airports consists of charter flights for package holidays. People are travelling in their thousands for two or three weeks of sun and sea. The traffic is non-essential and hence elastic. Some of this traffic also travels by the scheduled air services at special cheap fares. The sheer weight of this movement at peak holiday times creates problems of space along the approach roads and in the air, and both the aircraft and the traffic on the roads have adverse effects on the environment in terms of atmospheric pollution and noise intrusion. Should, then, the year-round needs, wishes and welfare of the residents on the ground be subservient to this apparently superior cause from which they may derive no personal benefit?

In a pattern for the future it is clear that the international scheduled air services must be maintained and developed as necessary. Suitable facilities must continue to be provided for them but without the package holiday traffic that would not necessarily be an immediate problem. Indeed, it might not be a problem at all. On the other hand, foreign traffic coming in to an airport also has its share of non-essential passengers in the shape of visiting tourists who perhaps are also on cheap packages. But there are not many countries that would wish to stop the influx of overseas visitors. They bring in foreign currency and bolster their trade and hence their economies. Some countries depend on the tourist business for their economic survival. Perhaps then the movement is not so 'non-essential' after all. It will certainly not be cast off lightly. Interestingly in this context, the government of the Seychelles – lush tropical islands in the Indian Ocean and a popular package holiday attraction – is considering the imposition of a £60 visitor's tax to 'help preserve the environment and improve facilities'.

This, then, is yet another dimension of the airport extension controversy. But to return to our pattern for the future: at this stage it can really only be recommended that each individual case be dealt with on its merits without any preconceived bias one way or the other. Endeavours must be made to balance the needs and dictates of all relevant interests and accommodate them with the inevitable compromises to the extent that it is physically practicable to do. But if present trends continue, it is inevitable that the time will come when at some airports and under certain circumstances, that will no longer be possible. When that time comes more positive policies will need to be formulated. The writer suggests that when and where conditions caused by too many people become untenable, it is the charter flights and sub-standard fares on the scheduled services that will have to be restricted and/or (on a selective basis) eliminated. But regardless of all else, the international scheduled flights of the world's airlines will have to be maintained to the extent that demand requires (maybe without special cheap fares and subject to no excessive provision due to wasteful competition) and proper provision must be made for them. But, remember always, planes pollute.

SHIPPING AND CANALS

Waterborne transport is not something that will figure largely in this suggested pattern for the future. The world's mercantile marine is here and here to stay and nobody would wish otherwise. Canals which are suitable as a transport artery (ie, they are of an adequate width and depth) have been constructed where it is practicable to do so and they are used as are natural rivers which have been made suitable for navigation. Furthermore, both the seagoing vessels and those on inland waterways are all reasonably environmentally friendly except for the fuel that they use. They serve an essential function in the conveyance of merchandise and the facilities that they provide should not be disturbed. The only control would relate to those of their activities which run contrary to the current environmental issues. These include such things as types of fuel in relation to atmospheric pollution, disposal of refuse in port and at sea, cleaning of oil tanks at sea and particularly oil spillage. However, these are ongoing issues which this particular suggested pattern need only endorse.

Having shown up water transport in a reasonably good light, the main consideration in this context is whether traffic could be transferred to it from the roads. In other words, there could be a case to move some traffic from roads to inland waterways and to coastal shipping where practicable and justified. This is something, therefore, that should be taken into account as and when and on its merits. It is something that the writer would like to see and it will be recalled from Chapter 2 that this is a view that is shared with the (British) Royal Commission on Environmental Pollution and with the Commission of the European Communities.

MOTIVE POWER

Motive power is not something that can take a prominent part in any one suggested pattern for the future. Local circumstances vary to the extent that a single line of thought becomes impracticable although there are two threads that run through everything. They are atmospheric pollution and a future shortage of oil. The former is something that must be (and already is being) taken into immediate account. The latter necessitates further research and development in the field of electric batteries which must continue. Only the first issue, therefore, can be commented upon here.

It is of course popular knowledge that electrically powered vehicles do not pollute the atmosphere in the course of their movement along the road. But the power stations that generate that electricity could if they are coal burning or to a lesser extent if they are oil-burning. The generation of hydro-electric power does not pollute and neither does nuclear power to any material extent – or does it? It does have other problems such as the disposal of radioactive waste which one brand of environmentalist abhors and it is beyond the scope of this work to deliberate on that issue.

It would seem clear that in environmental terms, where physical circumstances permit, hydro-electric power should be generated. It would

then be so used for transport purposes, including a predominance of electrified railways and, for urban transport, on the roads, trams and trolleybuses. But then again – are these ways of thinking really so clear? For
example, in 1997 China diverted the flow of the Yangtze river as a first
stage to building the Three Gorges dam, a project to harness the power of
the river and one that it is expected to complete by 2009. By that time it
will be turning the turbines needed to fuel China's modernization plans –
and without any gaseous emissions. And it will give better flood control.
And larger vessels will be able to navigate further upstream. But the people are losing their homes and they are not happy, even though they are
being compensated. And neither are the environmentalists even though
environmentally friendly electric power will eventually be generated in a
country which is destined to be one of the worst emitters of greenhouse
gases. True, the appearance of what is an area of tremendous natural
beauty may be altered but the antagonists also appear to be concerned
about possible silting, safety in times of war and the destruction of rare
fauna and flora. What is clear is that the environmentalists can have only
an input to the consideration of projects of this kind. They cannot rule.

At one time, in atmospheric pollution terms, the case was made for
trams and trolleybuses when consuming electricity generated also by oil-
fired power stations. But with the advent of cleaner diesel, the balance of
benefit in polluting terms is less. Furthermore, buses are now being
equipped with the 'Euro 2' engines which conform to EU standards on
emissions (which specify lower levels of particulates and polluting gases).
Many undertakings are now also switching to the ultra low sulphur fuel
(which is the highly refined fuel containing less sulphur than normal
diesel). It is not easy to measure the exact relationship between the pollutant properties of a bus so equipped with that of, for example, an electric
trolleybus, as the pollution from one is emitted along the public highway
amongst other traffic maybe in a busy urban area, whilst the other issues
from a remote power station with only a part being attributable to the trolleybus anyway. However, as a generalization, it is now reasonable to say
that there is little to choose in terms of atmospheric pollution (although
the mix may be different) between a diesel bus (with a Euro 2 engine and
ultra low sulphur fuel) and a trolleybus powered by electricity generated
from an oil or even a gas-fired power station. In pollution terms, therefore,
this must condition the thinking on the future of electric vehicles. With
nuclear or hydro-electric power, the situation would, of course, be different in terms of both pollution and the future availability of oil.

One further possibility arises. It is known that the worst effects of a
diesel-powered engine are experienced when accelerating and with constant changes of speed. If a diesel engine powered an on-board generator,
the electricity from which in turn powered the vehicle, the diesel engine

could then be set at a constant speed and fine-tuned for optimum performance and minimum exhaust. This would seem to be the ideal but the idea was the subject of trials in London (England) as far back as 1903 (albeit with petrol) and came to nothing. Nevertheless, in more recent years the principle has been widely adopted on the railways and its application to buses must, therefore, remain food for thought.

All of these aspects must continue to be the subject of research as will the development of alternative fuels. Hydrogen, for example, could be a possibility. But these are essentially technical issues and it would be neither sensible nor practicable to offer any quick arbitrary suggestions or predictions here.

Maybe the electric car will in due course become commonplace. People with mobile caravans generally use bottled gas for lighting and cooking. They have two cylinders and when one is exhausted they change over and at the first opportunity exchange the empty cylinder for a full one. Perhaps in due course an electric car will have two batteries and when one has no more life, then, like the gas cylinder and like changing the battery in a torch, exchange it (at the equivalent of a petrol station). At least it does seem reasonable to suppose that something will have been developed before the oil runs out.

IMPLEMENTATION

Let it be reiterated that this pattern for the future relates to broad policy for implementation worldwide, suitably adapted as necessary and when circumstances dictate. This means that it would be introduced in piecemeal fashion and although in some cases action is required now, it is inevitable that some of it would take time even with the best of intentions as, for example, back-door deliveries to shops. Many nations will be involved. Many people will be involved (and inconvenienced). Their accustomed way of life will be disturbed and, regardless of any urgency, that means delay. Nevertheless, it is hoped that readers will now think about these things and decide on the way that they want to go.

Remember again the fundamental difference between the provision of goods and passenger services which was mentioned in a preceding subsection. Many of the passenger services will not be profitable and even if revenue exceeds direct costs, for fixed track systems it may well not cover the extra capital costs of construction. In the case of underground railways it is virtually certain that it will not. For this reason, when dealing with financial issues, it is again the passenger side of the business that is mostly involved. Some expenditure on the goods side could nevertheless come if it was decided to modify or upgrade railway lines or improve or provide facilities for coastal shipping with a view to taking traffic off the roads.

The proposals as described are bold to the extent that they would be costly. A lot of money would be involved and money that will not be recouped from fares. At the same time, money is something that is always in short supply (in all countries) and whatever politicians might think, nobody is anxious to suffer heavier taxation. There is, of course, the substantial amount of revenue that would be forthcoming from fuel tax, as was discussed above, when considering a pricing mechanism; then there is always the prospect under some circumstances of the injection of private capital. But that apart, more would still need to come from public funds.

In the developed countries, items of heavy government expenditure are likely to include, *inter alia*, social security, welfare, health, education, transport, etc, all to greater or lesser extents according to national politics. A nation could live without high cost money guzzling social security payouts and a health service. Some individuals might not but the nation would and many nations do. Similar remarks apply to welfare and education (to take the examples quoted). They may be nice to have. Some individuals would probably suffer hardship if they were not there (some already do). But the nation (and the world) would continue to live if expenditure on these items was also something less than lavish (and again, some nations already do). On the other hand, an adequate transport system is crucial to the economic life of all communities and of all nations. Transport is one of the essentials of present-day living of the masses and the pulse must continue to beat. If it does not, then life as we know it today will cease. Allocations under the heads of public expenditure must, therefore, involve a consideration of priorities. It is not until that scenario is understood, recognized and accepted that proper transport networks will be provided and maintained. Although many will disagree and it may not be a popular view, it is the writer's contention that when it comes to priorities, transport, for the reasons stated, should be one of the first items for consideration. There will, of course, also be other essentials (such as, for example, the maintenance of law and order) which would also have to be taken into account. But then, with an economically sound and freely mobile nation, money that remains for state spending could then be allocated to the other 'luxury' or less crucial functions. The priorities must be right.

There is one other factor, however, in the implementation process to which attention must be drawn. The emphasis throughout this chapter has been on a return to public transport, at least in urban areas. But some people do not use public transport today because they do not like contact with their fellow passengers. There is an unruly element that behaves in a disorderly manner, soils seats, etc and damages equipment. When they are met, travel is at best unpleasant and at worst hazardous with a danger of assault. For this reason alone some people dislike public transport whilst others are too scared to use it, particularly at night, preferring the greater security of their private car. This is not something that is endemic in any one area. It has spread across the world from New York to London, to Sydney – or has it? The writer felt no qualms when riding around Singapore! The problem can, therefore, be tackled. But one thing is abundantly clear. Before there can ever be any worthwhile return to public transport, this issue will have to be properly addressed; although it is not within the powers of the transport operators to do it.

TEN

Epilogue

PREAMBLE

Having reached this final chapter, the first broad look into the complexities of the transport industry has been completed. But even just the principles of all of the many facets of transport of all of the modes in all of the countries of the world is a subject of exceptional magnitude; to condense that information into a relatively few pages must inevitably mean much remaining unsaid. As has been told, the difficulty is compounded by the fact that the world is now drifting towards undesirable and maybe impossible situations in terms of lack of space on the ground and far-reaching environmental damage. Transport is at the forefront of these problems and they have to be addressed. The facts and issues that have been discussed have therefore been drawn upon to suggest policies for the future and, however unpalatable, some of it in some places will have to be applied. To a large extent we are now proceeding into the unknown; perhaps more than ever before.

THE JOURNEY ONWARDS

The take-off

We were in our car and we were rapidly approaching a crossroad. There were traffic lights but they were neither red nor green. They were flashing amber like they do in Europe which means (appropriately) proceed at your peril! As is usual at crossroads, there was a choice of ways. There was straight on along our original route or there was a turn either to the right or to the left. And this had nothing to do with party politics. Whoever might influence our direction beyond the crossroads, indeed, of anything that might have an effect on this important issue which will influence our future way of life, must be the result of the deliberations of all-party and informed opinion unbiased by political preferences and popular votes. But, be that as it may and to continue with the parable, there were towns astride each of the three highways which lay ahead and, as has been said, we had a choice.

We decided not to change direction. Things seemed to be all right today and they will probably be all right tomorrow. We therefore kept to our original road and went straight ahead.

The first stop

We soon caught up with other cars, at first moving slowly and then blocking back in a queue stretching as far as the eye could see through to the centre of the town that we were approaching. Engines were still running; they had to be because every now and again there was a start/stop and through the poisonous haze from hundreds of exhaust pipes a solitary bus could be seen. The upper deck was clearly empty but there may have been a few passengers downstairs – the few who could not drive. The slow crawl lasted for several miles. The seats in the car were comfortable (at least for the first hour or so) and we were not cold. We had a heater which can either recirculate stale air within the interior or bring in fresh air from outside (which is what the car handbook said). However, being suitably polluted, to avoid choking and possibly worse, we had to shut it out. The town had had a by-pass for some years which was fine when it was built. However, it had quickly become overloaded and a super by-pass was provided more recently to by-pass the by-pass. But that is now also overloaded and offered little respite on this occasion.

We had hoped to see something of the town and maybe shop in the centre as in the end we did get there. However, there was little time left and it proved to be neither desirable nor even possible anyway. Parking, which was expensive and in short supply in relation to demand, involved queuing – and the queuing lane was full! Furthermore, the character of the town had been drowned by demolition to enable road reconstruction for traffic engineering schemes and for the provision of faceless multi-story car parks. Some of the shops had closed because the place had become so inaccessible. Reluctantly we decided to abandon this excursion. The pleasure had gone and that was before we had even begun to suffer from the effects of the poisoned air, let alone the predicted global warming.

We decided that perhaps, after all, it was time to change direction and after a further struggle we came to a roundabout. We returned to the crossroads.

The second stop

This time we turned the corner and took one of the alternative routes. We came upon another town but now one that was being governed differently. Environmentalists were in power. A block back was not encountered; in fact, looking at the road ahead there appeared to be not a lot of traffic. After further progression the reason became apparent: a sign indicated that entry of cars into the downtown area was heavily restricted. But a sizeable car park was to hand (as there were at other strategic points on the periphery of the built-up area). For a reasonable combined parking and travel charge (and much cheaper than the cost of petrol), a service of double-deck buses conveyed people into the centre and we took advantage of that facility.

So far, so good. The use of what was quite a narrow road (but also with ample footpath and cycling accommodation and a row of small trees) was maximized by the use of large capacity vehicles with passengers all comfortably seated. As a cleaner variety of diesel fuel was obligatory on what were well-maintained vehicles (plus, with the large vehicles, a minimum amount of exhaust for the maximum number of people), the amount of any damaging effluents was not great.

Having arrived by bus there were, of course, no parking problems and it was refreshing to be able to walk around. First impressions were that there had been some excellent town planning here. The original charm of the central area had been revived as the need to cater for an excess of vehicular traffic did not arise. Strategically sited footpaths and bicycle tracks allowed pedestrians and cyclists to go their way without let or hindrance by vehicular traffic and the landscaped areas were green and inviting. Birds were chirping because their natural habitats had been restored and even a local stream now had fish. The air was fresh and clean as with few cars, most of the poisonous emissions had disappeared. Cycling and walking does not pollute and what ill effects there were came mainly from being blown across from the town which we had recently visited.

We noticed that the shops in the centre were all open and flourishing. Prices, however, were higher as the more stringent regulations and timings which now governed wholesale deliveries had increased trading costs. We saw people emerging from a supermarket with nicely laden trolleys but they did seem to have a problem in loading the contents on to their bicycles! They would have to hurry otherwise their frozen food will melt. They could have gone to the nearest bus stop but that might not have been easy either and walking was really out of the question. They could perhaps even have used the tram. A local light rail network had recently been opened to strong public acclaim. It was pollution-free on the ground and the power station which was producing its electricity was many miles away. Somebody else could worry about that. But as light rail is expensive not many routes had been constructed. It might not, therefore, have been convenient for the people that we saw even though we understood that its provision was being painfully reflected in their local taxes. Nevertheless, it was good for the few who did actually live near a tram stop. It was just a passing thought at the time but, with the high cost of the vehicles and related infrastructure, maybe new buses for the entire area could have been provided for the same money, perhaps several times over, and then everybody would have had the benefit.

We strolled on a little further to where children were coming out of school. The footpaths and cycle tracks would now really come into their own. At last children could move around without the need to cross busy roads with the ever-present fear of accidents. Even so, we did ourselves experience a dangerously near miss with a bicycle that we did not hear coming and which had little regard for our presence. That is certainly a hazard. In some cases cyclists and pedestrians were segregated with appropriate signs but few cyclists worried about that and claimed priority everywhere regardless of any foot traffic. We were surprised nevertheless to see parents standing at the school gates (equipped, incidentally, with macks and umbrellas as the sun had now gone). It transpired that although the ill effects and dangers of heavy traffic had indeed been eliminated, the less desirable members of the community had not. Children walking or cycling along quiet ways through attractively landscaped grassed areas and bushes, were in danger of being molested. In fact, the mothers did not like it too much either and they themselves arranged to come in groups. In darkness no way would any of them venture through to the shops or evening activities in town; the pedestrianized underpasses and walkways had become virtual no-go areas. We left them to face what was now to be a cold and wet walk home whilst we returned to our park-and-ride bus and thence our car, both of which were dry and warm.

Returning to the car park proved to be no problem and our car had not been vandalized. The ills of the first town that we visited had certainly been overcome in the second. But we never did see how those laden trolleys from the supermarket had been dealt with and all in all it was still far from Utopia. The weather had now turned sour and we were glad that we were not destined to exchange the warmth and comfort of our car for a bicycle saddle.

Back, therefore, to the crossroads.

Where now?

There is only one way left to go. And what now can possibly lie before us? Through the chapters of this book the facts have been considered and the issues discussed, whilst bearing in mind the environmental effects.

The time is, therefore, opportune to think about what we would wish to see in this third town that we are now approaching. But more than that: we must also think about what should happen in the field of total transport all over the world, now, on the brink of the twenty-first century, and then over the years beyond that.

We have now taken the third road and there is another haze in the distance. It is not pollution this time but we do not know what it is and neither do we really know what we expect and hope to see. Before proceeding too far, therefore, we must think about how and why all this has happened as it might help to crystallize in our minds any future patterns and policies. We will stop and cogitate, mull over the background to it all and consider what has already been said.

We pull in to a lay-by.

AN OVERVIEW

Railways came into being in the 1830s, less than 170 years ago, but can now reach a speed of 220mph (354kph).[1] Movement by air as we know it today and which has revolutionized long-distance global travel, with aircraft that can fly more than twice the speed of sound or capable of carrying some 400–500 passengers, has developed over a span of only some 50 years. But even the railway's period of gestation is infinitesimal – nothing at all when compared with the history of the earth and its period of evolution. In other words, and transport is not unique in this respect, things are now moving rapidly; so rapidly that the realities and effects have come before there has been time to anticipate them. As was said in the Introduction, parts of textbooks can become out-of-date before they are printed.

Much environmental destruction is being caused today, wilfully or otherwise, but it is an inevitable product of human endeavour. There is little that can easily be avoided and probably nothing without anguish. But the damage can be only partially attributed to transport. In Great Britain, for example, some 74 per cent of the poisonous effluents emanate from industry. That attributable to transport is depicted in Table 10.1. The problems must be addressed regardless of their source but this book is concerned only with transport and its effects; it is the transport-related issues which are relevant to this study. Whatever may be the solutions or the palliatives, it is the national governments that will have responsibility for introduction and enforcement within their respective territorial boundaries. Having said that, one of the more important concerns of the day is the pollution of the air and the oceans but physically, if not politically, the atmosphere is international as also is the sea. Whilst, therefore, considerations may be conducted independently by different nations, if they are to

Table 10.1 Sources of Carbon Dioxide Emissions in Great Britain.

	%
Road	22
Railway	1+
Civil Aviation	1+
Shipping	1+
All transport	26
Non transport	74
Total	100

Source: Transport Statistics. 1997.
HMSO (GB) from National
Environmental Technology Centre

succeed, application must be internationally enforced and this could become as complex and as difficult a task as it will be to find the remedies. This is something that might well be addressed by the various international organizations, the relevant bodies having already been described in Chapter 7, and they do have the environment on their agendas.

If development of any kind offers some new and advantageous facility without too many damaging side-effects, then all well and good. But it is not always like that. Throughout this epistle references have been made to the environment and its abuse. Chapter 8 dwelt on the environmental issues and the dire consequences if suitable action is not taken soon; and soon really means now. But if we are not careful it will be like running after an express train after it has sped through the station and a red signal whilst rushing forward to its ultimate doom. Even now it may not be easy to catch it in time. It may not be easy because neither will it be popular if the environmental issues are properly dealt with. But it is the responsibility of governments to point people in the right direction. As the destruction of the environment is a worldwide process, for success, every government of every nation must play its part. Coercion in its strongest form will need to be exercised. Amelioration of the impending damage can therefore lie only in the hands of politicians and then comes another problem. Having entered this arena, other objectives come to mind which the politician might also see as important. Some politicians do not always have the right priorities. Sometimes their priorities are as below, in descending order:

1. Power
2. Politics
3. Policies
4. People

Votes come first – people come last.

Basically, the environment is a political issue but how high is it on the political agenda? As a generalization, the prime objective of a politician is to get power and into government (the writer has no particular country in mind). There are three ways of doing this:

- to be elected in a democratic manner by high pressure salesmanship with doubtful promises made to a fickle-minded and gullible electorate whose primary thought is 'what is in this for me?' (like the public consultation process discussed in Chapter 6);
- to be 'elected' through ballot rigging or with just one candidate;
- to achieve power by force with bloodshed if necessary.

It is not difficult to think of examples where governments have come to power through each of these means.

Having come to power, then, and maybe regardless of any manifestos, the fundamental principles of their particular brand of politics – their 'sacred cows' – must be implemented, not even modified, regardless of the circumstances of their application. Last of all comes the interests of the people, the nation and the world. Small wonder that environmental deterioration will be a difficult nut to crack. But it is not a difficulty that is entirely of their making. If politicians campaigned for, say, severe car restraints, a ban on cheap tourism and a compulsory move to public transport, just to quote examples, they would most likely not only lose their seats, they would lose their deposits as well! They would have about as much chance of gaining power as there is of walking to the North Pole in a bikini. And the door would be open for another party eager to govern and with a conveniently less repressive style. There are some able people in governments, both in power and in opposition, but little progress will be made while they are required to use their talents and their energy in fighting each other just to stay in office, rather than in addressing the serious and unpopular issues that must be faced. But more than that, it has been said that every country, worldwide, must participate in the protection of the environment if such actions are to be fully effective. This includes the governments which have come into being through the second and third alternatives listed above. No, it will not be easy.

As this is not a book about political science, it will just have to be assumed that governments worldwide will somehow, perhaps through some sort of coalition, get themselves into a position whereby they can implement to the full extent whatever might be necessary to alleviate the environmental hazards discussed in Chapter 8. There are also the international organizations which can hopefully bring pressure to bear. But however difficult or impossible it might be, it is still just one hurdle. It remains for decision what needs to be done. Every individual will be required to play a part as must industry generally. Transport is one of those industries and in this respect it is a major player. The causes of the problems have been identified but those causes often constitute a desirable part of our daily activities. The medicine is likely to be unpalatable and even if it was not as drastic as all that, it would still cause public displeasure. The common thread is that we want to maintain our quality of life. But that must include preserving the environment. There cannot be one without the other. But if we want to preserve the environment it means not doing some of those things that we want to do and what we now accept as part of the quality of life which we want to keep. And therein lies the dilemma! We live among opposing forces. The way forward will not be easy.

When thinking about these things it must be recognized that the way of life between people in the developed and parts of the developing countries is different – sometimes very different – even though the effects of pollution and the destruction of natural habitats spread worldwide. Some of what has been said has been directed at the now accepted activities of people domiciled in the 'Westernized' parts of the world. Clearly it would be impracticable to offer a common set of guidelines for all to follow. The tribal head in his straw hut in a clearing in an equatorial forest and who survives with the aid of a bow and arrow does not necessarily take his car to an out-town supermarket. Neither may he be damaging the environment. But his way of life may be endangered by those who are. Possible courses of action have been floated in Chapter 9 but no, it will not be easy.

Perhaps after all the haze ahead may lift. Perhaps it will all blow away. On the other hand, maybe that is wishful thinking, although it might make life in the immediate future a little more congenial. But conditions in the first town that we visited were unacceptable. We could not live there nor even make a quick visit. But looking at the information and statistics as they have been presented, who can dispute that they are the conditions which are destined to come? Then there was the second town. All very clinical and orderly and free from poisonous fumes. But it was apparent even from just that short visit that without their cars, too many people were experiencing too many problems. Here the haze had blown away but the resultant exposure left certain things still to be desired.

THE ARRIVAL

The product of all transport is a safe arrival and we are now about to arrive. But we are not yet quite at the end of the road and there are divided views on what we hope we are soon to see. At the moment we can only fantasize. However, unlike previous editions of this work, the final paragraphs of this edition cannot rest with just a few words about oil supplies running out and a resurrection of trams and trolleybuses.

This time readers must fantasize for themselves and then develop their own visions into possible lines of concerted action. In so doing and having been acquainted with the facts, they might like to take into account some of the thoughts that have been expressed in these latter pages. On the other hand, they may have different ideas. What is to be done about traffic congestion? What is to be done about pollution? Both are issues of extreme importance and both strike at the heart of the principles of transport. But whatever those ideas are, it is the students of today that will determine the happenings of tomorrow and it is the readers who will now write for themselves about the future that they hope to see.

References

Chapter 1

1 P&O Annual Report and Accounts 1997.
2 The Trans-European High Speed Network. The Field of Railway Exellence. UIC 1997.
3 Bulletin of the European Communities – Supplement 3/93.
4 The Citizens' Network. European Commission. 1996.
5 Six-Year Transit Development Plan for 1996–2001. King County Transit Division, Seattle.
6 London Transport Statement of Strategy 1997.

Chapter 2

1 Basic Road Statistics, 1997. British Road Federation.
2 Collective Transport in the Brussels-Capital Region. STIB 1996.
3 Transport and the Environment – The Royal Commission on Environmental Pollution's Report. Oxford University Press, 1995.

Chapter 5

1 State Transit Corporate Plan 1997/8 – 2001/2. State Transit, Sydney.

Chapter 8

1 The Coming Plague: Newly Emerging Diseases in a World out of Balance. Virago.
2 Occasional Paper 10. Greenhouse Gas Emissions in Australian Transport in 1900 and 2000. Bureau of Transport and Communications Economics, Canberra. 1995. Dr Leo Dobes.
3 Bulletin of the European Communities – Supplement 3/93.
4 *Ibid.*

Chapter 9

1 Road Reform – The Way Forward. Roading Advisory Group, 1997. Government of New Zealand.
2 EC Transport Policy and the Environment – an Overview. Institute for European Environmental Policy, London. 1994.
3 Transport Statistics. 1997. HMSO (GB). From 'National Accounts' (OECD).

Chapter 10

1 A Thalys train between Brussels and Paris. 'Panorama' 4th quarter. 1997. UIC.

Index

Note: If consecutive pages have relevance only the first page number is shown.